# MANAGING CONSTRUCTION CONTRACTS

# MANAGING CONSTRUCTION CONTRACTS

## Operational Controls
## for Commercial Risks

ROBERT D. GILBREATH

A Wiley-Interscience Publication

JOHN WILEY & SONS

New York • Chichester • Brisbane • Toronto • Singapore

*Library of Congress Cataloging in Publication Data*

Gilbreath, Robert D. (Robert Dean), 1949-
  Managing construction contracts.

  (Construction management and engineering, ISSN 0193-9750)
  "A Wiley-Interscience publication."
  Includes index.
  1. Building—Contracts and specifications.    I. Title.    II. Series

TH425.G53 1983      624'.068      82-15983
ISBN 0-471-87635-6

Printed in the United States of America

10  9  8  7  6  5

For Linda, Bob, and Alice,
who met me at airports,
delivered me back to them,
and listened to the
sound of my typewriter
in between.

# SERIES PREFACE

Industry observers agree that most construction practitioners do not fully exploit the state of the art. We concur in this general observation. Further, we have acted by directing this series of works on Construction Management and Engineering to the continuing education and reference needs of today's practitioners.

Our design is inspired by the burgeoning technologies of systems engineering, modern management, information systems, and industrial engineering. We believe that the latest developments in these areas will serve to close the state of the art gap if they are astutely considered by management and knowledgeably applied in operations with personnel, equipment, and materials.

When considering the pressures and constraints of the world economic environment, we recognize an increasing trend toward large-scale operations and greater complexity in the construction product. To improve productivity and maintain acceptable performance standards, today's construction practitioner must broaden his concept of innovation and seek to achieve excellence through knowledgeable utilization of the resources. Therefore our focus is on skills and disciplines that support productivity, quality, and optimization in all aspects of the total facility acquisition process and at all levels of the management hierarchy.

We distinctly believe our perspective to be aligned with current trends and charges that portend the future of the construction industry. The books in this series should serve particularly well as textbooks at the graduate and senior undergraduate level in a university construction curriculum or continuing education program.

JOHN F. PEEL BRAHTZ

*La Jolla, California*
*February 1977*

# PREFACE

Commercial pressures of the present and forecasted construction environment have caused project owners and contractors to reexamine the ways in which they contract. When faced with a construction landscape dominated not only by massive and complex technical challenges but increasingly menacing legal, financial, and business risks, those responsible for project success face a choice. They must either employ outdated processes and failed management techniques or accept the challenge by planning for and using sound business controls commensurate with the nature and magnitude of the risks they face. This book is dedicated to those who choose the latter course.

The risks of contract selling and buying are obvious. They include the seemingly omnipresent specter of litigation, disputes, schedule delay, and a sea of changes in the scope, duration, and cost of construction work. Often the magnificent technical achievements demonstrated by the completion of a construction project, and its improvement of the lives of people it benefits, are obscured by the commercial difficulties and entanglements it entails. Project completion should be a time of rejoicing and satisfaction for all participants rather than a minor signpost on the road to financial or legal confrontation or ruin.

Commercial risks were always present in the past, yet they were secondary in importance to a much larger technical challenge. Why is it then that today project managers, their staffs, and corporate executives speak of a project in terms of lawsuits, claims, change orders, and backcharges rather than new production capacity, increased profit margin, or advances in environmental and social conditions? The answer lies in the growth of commercial risks to project success and the inability of traditional management techniques and tools to recognize, avoid, and control these risks.

This is not to suggest that managements are insensitive to commercial risks or unmotivated to counter them with the best controls available. Too often, however, attention is focused on the symptoms of commercial failure once it has occurred, not on eliminating the potential for failure before it occurs. As in other fields, prevention is sometimes overlooked until damage is felt. As any fire marshal will tell us, fire prevention is much more cost-effective than fire fighting or the repair of fire damage. And physicians universally agree that preventative medicine is much preferable to the treatment of a disease once it has been contracted.

This book follows such an approach. It presents effective means and methods to control the risks of construction contract efforts—preventative medicine rather than postmortems. Of course, problems must also be discussed in order to determine corrective action for future benefit. One must learn from past mistakes in order to prevent their reoccurrence. To ignore these lessons, or to fail to plan and employ controls to counter specific risks associated with particular projects, would be abdicating one's responsibility to assure project success, no matter the organization or the level of participation.

How then does this book help project participants address commercial risks to the contracting process? It does so by first describing the environment in which construction contract business takes place. Alternative organizational settings and the roles and relationships of all project participants are explored. The four major phases of contracting are then presented, with detailed explanations of the business objectives, common risks, and proven controls for each phase.

Typical activities of major participants are described for each element of contract management. Sample forms, documents, reports, and records are included not only to demonstrate their use in an effective management program but to allow for rapid tailoring and application to actual contracts in the planning or execution stage. The role of contract documents themselves is defined, as well as considerations for their selection, use, and enforcement. Alternative pricing strategies are presented, together with recommendations for achieving control objectives through selective application. The need for forward-looking contract planning (when and how to employ contracts) as part of an overall project plan is also emphasized.

The effective management of construction contracts begins long before the documents are signed. Risks and errors, though latent during the enthusiastic stage of project commencement, almost always surface and impede the project once construction has begun and plague it throughout its life. For this reason, the first half of the book explores planning and formation controls, that is, how does one achieve the best contract, under the best terms, with the best contracting parties and at the best possible price?

The remainder focuses on the management of an active contract or contracts once construction has begun. Common and not-so-common commercial transactions are presented, with the ultimate objective of quality work, on time and at the agreed-to price, guiding each specific contract administration task. Ways to ensure proper progress payment and the objective resolution of changes, as well as the pursuit or defense of claims, are discussed. Streamlined control procedures and easy-to-use forms and reports are given in order to assist those performing and managing these contract administration tasks. And, finally, a management perspective is taken to present ways in which proper contracting practices can be assured. The use of contract personnel, contract reporting, and contract auditing as part of a stable contract management program is discussed.

The material, recommendations, and suggestions contained in this book were drawn from hundreds of projects throughout the world. It is hoped that they will assist the owner, design professional, and contractor in focusing attention on commercial risks, reduce the incidence of commercial failures, and direct their management skills and talents toward their joint and primary objective, successful project completion. Should this be the case, the book's objective will be satisfied many times over.

ROBERT D. GILBREATH

*Norcross, Georgia*
*December 1982*

# CONTENTS

# LIST OF FIGURES

# LIST OF EXHIBITS

# MANAGING CONSTRUCTION CONTRACTS

# INTRODUCTION

Major construction projects represent some of the largest and most complex undertakings known. When completed, they give testimony to a variety of technological methods, engineering skills and innovations, materials, and a tremendous consumption of required resources—time, money, and people's talents. What is often overlooked when viewing a massive structure or an intricate operating facility upon its completion is the intricate web of commercial transactions, business arrangements, and management challenges that were also required.

Much has been written, proposed, and implemented in the way of management controls over the construction of major projects. These efforts involve attempts to control the performance of the project as it progresses — performance in terms of cost, schedule, and technical aspects. And yet projects continue to fail. They fail in terms of cost performance when budgets must be constantly revised upward and when final cost greatly exceeds original estimates. They fail in terms of schedule when completion dates are extended and when the completed facility is available for use long after its intended availability date. And they fail when the completed facility, or portions thereof, do not meet specified technical requirements or are otherwise unfit for their intended purpose.

The business of construction is performed through contracts. And from the perspectives of both owner and contractor it is a business rapidly becoming fraught with risk and uncertainty. The impact of project failures has caused owners and contractors to reexamine the means and methods by which they contract for engineering and construction services. They are searching for ways to improve the processes by which they plan, form, and administer construction contracts and to control commercial risks that threaten project success. This book is intended to help them in that search.

1

## COMMERCIAL PRESSURES IN THE CONTRACTING ENVIRONMENT

Today's construction environment is fast becoming more perilous then ever before. Ever-present cost and schedule overruns, pervasive construction disputes and claims, and the abundance of change orders associated with many projects are symptoms of an inability to control or mitigate commercial risks. The frequency and intensity of these risks are increasing because of a number of factors:

Growing commercial pressures caused by inflation, resource scarcity, financing costs, and more stringent public accountability

Closer outside scrutiny of management controls and practices

Proliferation of design, construction, and management alternatives

Increasing involvement of governmental and regulatory entities

More frequent use of joint ventures, pooled ownership, and other project-specific corporate combinations

Growth of innovative and complex risk-sharing techniques

Increasing complexity of today's contracts

## OPERATIONAL CONTROLS

Rather than focusing on postfailure analyses, many project owners now concentrate on ways to control performance before failure occurs. Their attention has shifted from review of damaging claims and litigation to the sources of these contractual failures. And they are designing and implementing "operational" controls over the contracting process — preventative medicine rather than postmortems. An emphasis on operational controls requires an analysis of each major element of contracting and an active desire to increase its efficiency and effectiveness.

## CONTRACTING ELEMENTS

For owners, the challenge of contractual controls may be phrased in the following question: How do I achieve a fair and equitable contract that provides acceptable levels of control, and how can it best be administered throughout the performance period? The answer to this question lies with the recognition that contracting controls are vital to project success and that the stakes are too high to ignore time-proven methods ensuring the owner's protection during all phases of the contracting process.

The contracting process begins with the inception of a need to contract for outside services and a determination of the scope of work and continues through the completion of contracted performance, final payment and acceptance, and formal contract closeout or termination. Whether an owner contemplates a single design–build contract or as many as 100 prime contracts, each step contains potential for control or failure.

For each contract, four major elements or phases demand management attention and control (see Figure 1):

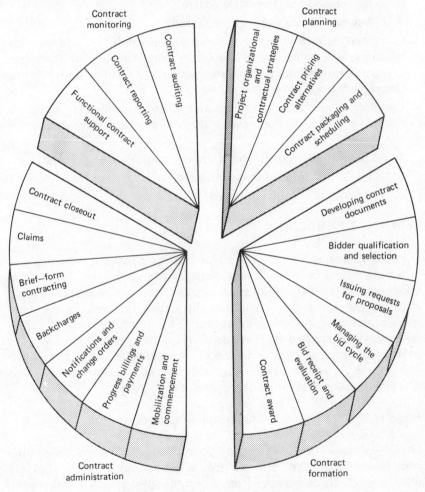

**FIGURE 1.** Major phases of contracting.

*Contract Planning.* Contracting practices should be planned and organized in a manner consistent with

1. Overall company objectives
2. Project-specific circumstances
3. Technical, schedule, and commercial risks

*Contract Formation.* Construction contracts should be entered into only when the owner is assured that

1. They are fair and equitable
2. They contain provisions that protect the owner's interests
3. They embody performance controls
4. They can be properly administered.

*Contract Administration.* Adequate procedures, enforceable administration controls, and trained personnel should allow the owner to control, rather than react to, commercial risks and pressures during the performance of contracted work.

*Contract Monitoring.* Owner resources and management attention should be focused on the previous three contracting elements through the use of:

1. Adequate functional support to each contracting process
2. Timely and informative contract reporting
3. Cost-effective contract auditing

## CONTRACT PLANNING

Because the construction of a major facility is usually secondary to other ongoing owner operations, such as the production and sale of goods or services, owners are generally at a decided disadvantage when entering the world of construction contracting. Before doing so, there is no substitute for thorough planning. This planning should follow a structured, disciplined approach considering the owner's objectives, the existing contracting environment, and alternatives for project execution and control. As such, an owner's contract plan should address the details described in the following subsections.

### Project Organizational and Contractual Strategies

Owners should choose an organizational alternative that best suits their control objectives and requires a level of involvement they are able to meet. Options range from a design–build approach, where the owner awards one huge contract for virtually all project-related services, to the multiple – prime concept, where as many as 200 separate contracts will be awarded. In between these two extremes lie a wide variety of approaches, each requiring different control strategies and different levels of owner involvement throughout the project's life. Owners should review each possible approach, analyze its features and benefits, and select the one best suiting their objectives and capabilities.

## Contract Pricing Alternatives

There are an infinite number of ways to price and pay for contractual performance, ranging from the straightforward lump-sum method to highly sophisticated cost-sharing techniques. To each may be applied a myriad of cost, schedule, and technical incentives, escalation provisions, and risk-sharing elements. Each has been effective given certain project-specific circumstances, and each has failed given others. The selection of a specific pricing method should be based on a thorough understanding of its features, control benefits, and the environment to which it will be applied.

## Contract Packaging and Scheduling

The subdivision of work into contract or subcontract packages and the sequencing of their formation and award activities over time represent a major challenge to owners who choose to award more than one contract for a project. Often contract documents are rushed through the formation cycle without adequate definition of scope, complete design information, and due consideration for project schedule needs. The result is uncompetitive and overpriced bids, overlaps and gaps in scope of work among several contracts, and an uncoordinated contracting program. Thorough planning and scheduling of each contract process is needed, along with adequate scope definition and a tailored pricing scheme. Well-conceived contract plans should be implemented before bidding commences, and critical schedule constraints in the design and construction phases should be incorporated into the contract bidding and award plan.

Such formalized contract planning is valuable to an owner in many ways. It surfaces scope inconsistencies among the anticipated contracts, affords conscientious pricing decisions, and outlines the support information, from the engineering and design efforts, required for contract bidding and award. It helps those responsible for contract formation itself to organize and plan their functions and allows owner construction management personnel to load required field monitoring resources and support services over the construction period.

## CONTRACT FORMATION

The process of contract formation begins with the decision to contract for project-specific goods and/or services and ends with a written agreement signed by both parties—the contract. Poor bidding and contract documents, inadequate bidder qualification and bid management, and poor negotiation efforts are damaging in themselves, but they are equally damaging to the contract performance and administration efforts that follow. They result in selection of the wrong contractor, often at the wrong price, and under the wrong terms. Owner control over the formation process involves the steps discussed in the following subsections.

### Developing Contract Documents

If you want it, get it in the contract: this advice has never been more warranted than it is today. A good set of contract documents is one in which the elements of the contract are clear and enforceable, the intent of the contracting parties is without question, and the interests of the owner are well protected. Most owners spend considerable effort in designing and adapting "reference" or "standard" contract documents. They should not only represent the highest standards of legal practice but also reflect sound engineering and construction judgment and include effective management controls. The use of standard or reference documents, each tailored to a particular application, lessens the possibility of misunderstanding, undue compensation, the likelihood of change orders, and the occurrence of claims or litigation arising out of contractual performance.

### Bidder Qualification and Selection

Owners who are not restricted by regulations or funding requirements to the contrary are often free to choose prospective contractors from whom to receive bids. These prospective contractors should be prequalified to rigid quality, financial, and performance standards before being allowed to submit bids. Proper bidder qualification helps to reduce the bid evaluation effort by eliminating those prospective bidders who prove unacceptable. It helps ensure contractor stability and therefore avoids potential performance problems should an unqualified contractor receive an award. And it promotes the widest solicitation of qualified bids, thereby increasing competition among bidders and driving the anticipated price downward.

## Issuance of Requests for Proposals

When preparing the contract bid requests, owners can protect their interests in two ways. First, they can disseminate all available information to bidders, allowing for competitive, precise bidding based upon facts rather than upon assumptions or contingency considerations. Secondly, they can channel the incoming bids into a structure that facilitates speedy bid evaluation, reduces postbid clarifications and negotiations, and ensures that owner terms and conditions are incorporated into the resulting contract documents.

## Management of the Bid Cycle

The selection of the best contractor under the best terms and conditions and at the right price begins long before bids are received and evaluated. During the bid cycle—the time from release of requests for proposals to prospective contractors until bids are received—owners can take steps that affect the quality and price of bids to be received. These include the judicious use of clear and timely modifications to the requests for proposals (addenda), adequate site visits, should they be required, and controlled and coordinated prebid meetings. Each of these steps is designed to give bidders clear and consistent information and allow them to estimate prices precisely—without unwarranted padding for perceived risk and contingency.

## Bid Receipt and Evaluation

Once bids are received by the owner, two interrelated objectives should be met: (1) a complete and impartial review of each should lead to selection of the best offer; and (2) the confidentiality of bid and bid evaluation information should be maintained. When evaluating received bids, owners should consider not only first costs, that is, the cost to have the work performed, but also the costs of operating and maintaining the finished work in the future. These "life-cycle" costs often result in the owner selecting a bidder who, though not the lowest, will provide the least expensive product in the long term.

Bid security measures should be initiated before bids are received and followed throughout the evaluation process. Failure to secure confidential information may result in reluctance of contractors to bid on future work (thereby reducing competition), compromised negotiation postures, and the commencement of a contractual arrangement on a less than good faith basis.

## Contract Award

As soon as a successful bidder has been chosen for award, the contract documents that will govern the arrangement should be prepared and executed by both parties. This generally involves the incorporation of bidder-specific information (prices, schedules, and so on) into the contract documents, as well as the accommodation of all addenda issued during the bid cycle. Quite often other postbid agreements are also documented in the contract, such as the results of clarifications and negotiations. It is not uncommon for owners to delay this process and instruct the contractor to commence work under "letters of intent" or similar agreements, pending conformance of the actual contract documents. Letters of intent have a limited use and a questionable value when used in this regard. They typically lack performance controls and leave the owner and contractor exposed to commercial risks.

## CONTRACT ADMINISTRATION

*Contract administration* is a term describing the commercial handling of a contract once it is awarded and until it is formally terminated (contract closeout) or dies an unplanned death (through contract default or early termination). Contract administration controls are designed to ensure commerical compliance with the terms of each contract. Even under ideal conditions, this is generally an arduous task for project owners. They must monitor not only the compliance by contractors but also their own compliance. Routine duties involving receipt and review of contractor submittals, maintenance of voluminous contract records, and monthly progress payments sometimes obscure critical risks encountered during the performance period. Change orders, claims, early payment or overpayment, and unsatisfactory performance can be prevented by implementing a structured contract administration program.

### Mobilization and Commencement

The initial phases of contract administration set the stage for its success or failure. Steps taken to begin a good faith relationship, one that will prevail throughout the performance period, include formal commencement meetings between both parties covering the commercial transactions they may encounter. In addition, this period requires that owners establish complete, protected, and easily accessed contract files. The mountain of contractual records, documents, forms, and reports needed for proper contract administration requires careful planning and accommodation. On major projects spanning several years, and given the transient nature of personnel assignments, this documentation becomes even more critical.

## Progress Billings and Payments

Unlike the purchase of finished goods, the buying of contract services requires periodic progress payments as the work evolves. Unplanned or subjective determination of progress results in overpayments to contractors. This can damage the owner in two ways: in view of today's cost of financing, owners who pay for performance early incur unnecessary costs due to the time value of money; and once payment has been made for performance yet to be accomplished, owners lose the payment leverage needed to ensure proper and timely completion.

## Notifications and Change Orders

Virtually every contract will change to some extent before it is completed. Formal and constructive changes represent the most pervasive and threatening factors jeopardizing successful project completion. A change-control program that identifies changes early, allows for thorough evaluation of the need for a change before it is ordered, and provides responsible pricing and payment controls is essential for project owners. Changes should be controlled on a selective basis that allows management flexibility to accommodate unforeseen circumstances and yet provides a minimum degree of control at higher levels. Balancing the risk of improper change authorization with the freedom to implement management decisions in a timely manner is a major challenge to project owners and their agents. Once authorized, changes should be monitored and controlled in the same manner as are original contract scopes of work.

## Backcharges

The practice of charging a contractor for the cost of correcting inadequate work or work that was not performed represents a reversal of normal owner–contractor roles. Many owners fail to collect these costs by not notifying the offending contractor or vendor before undertaking the subject work, or by inadequate control of costs or cost records. As a result, they often pay two or three times for completion of a single element of work. Aggressive backcharging, where appropriate, can represent substantial savings to owners in this position. And the incidence of backcharge situations increases as the number of prime contractors and vendors associated with a construction project increases.

## Brief-Form Contracting

Most sophisticated owners realize that all contracting needs and transactions cannot be anticipated months or years in advance and incorporated into contractual provisions. For this reason, they allow certain efforts to be accomplished on a local, or project-site, level. Generally, these involve the use of abbreviated or short-form contracts for short-duration, low-cost work and field-office-issued change orders in order to accommodate emergency or unforeseen events. Judicious use of these methods can represent a sound control policy that considers the need for responsive reaction to events that inevitably occur during the construction period. When these methods are used, however, the necessary control over contract formation and administration should not be abandoned for the sake of brevity or expediency.

## Claims

Most construction claims are not anticipated until they are received or presented from one party to the other. Although no two claims are alike, they do follow general characteristics, stem from similar circumstances, and have predictable elements. Owners embarking on construction efforts should recognize these facts and prepare for the defense or pursuit of claims. The best claims defense will always remain the avoidance of claim-provoking conditions through proper contract formation and administration. When claims are unavoidable, every effort should be made to identify claims situations early, thoroughly analyze each claim, and press for rapid resolution. Prolonged disputes or claims jeopardize remaining project performance and create unnecessary and bothersome litigation, generally to the detriment of all parties involved.

## Contract Closeout

Whether terminated in a planned or unplanned fashion, each contract should be formally "closed out" when future performance is no longer required. The process of contract closeout should be structured to protect the owner's interest by ensuring: (1) that all performance is acceptable (or otherwise noted); (2) that all expected documentation is received and adequate; (3) that any other deliverables have been received; and (4) that no damaging postperformance actions will take place. This process involves the joint effort of many people on the part of the owner, and their efforts should be coordinated to ensure that final payment is not made while any element of performance is lacking. In addition, information regarding each contractor's performance should be relayed to those responsible for qualifying potential contractors for future work.

## CONTRACT MONITORING

There are several tools available to project owners that help ensure that the controls needed for proper contract planning, formation, and administration are in place and operating in the owner's interest. Of primary importance are the organization, responsibilities, and personnel supporting the contracting process. Internal controls that owners should consider include informative contract status and performance reporting and contract auditing.

### Functional Contract Support

The assignment of contract formation and administration responsibilities to owner organizations may greatly affect the success of these duties. Many owners have wrestled with the issue of centralization of contracting processes versus decentralization, as well as the proper organizational position and reporting relationships of a contracting support group. Policy and procedures should be established for these efforts, and they generally require trade-off decisions between the owner's desire to manage the contracting process with the greatest amount of control and consistency and the need to afford operating groups the freedom to respond to the need for contracting in a manner that best suits their particular interests.

### Contract Reporting

Owner management should be apprised of contracting status and performance information on both a continuous and exception basis. Periodic management reports should be issued to identify key factors influencing successful project completion. Contract reports identify and describe such threatening elements as claims, change orders, backcharges, default, as well as cash requirements to support the construction effort. A system of contract reporting should provide for exception reporting and allow traceability of problems to their source—thereby allowing management to evaluate and take corrective action in a timely manner.

## Contract Auditing

Two distinct types of contract auditing assist owners in ensuring that their interests are being protected. Cost audits allow owners to verify that charges made by contractors, and subsequent payments by the owner, are accurate, in accordance with the contract provisions, and represent work that was actually performed. The audit approach and work program should be tailored to each specific contract type and focus on major elements or risks regarding error or overcharge. Operational audits, on the other hand, are forward looking in perspective. They seek to give owner management insight into the efficiency and economy of contracting practices. When properly structured and targeted, operational audits can help eliminate unnecessary risk, streamline operations, and ensure consistency of contractual controls.

## BENEFITS OF CONTRACT MANAGEMENT CONTROLS

The controls described in this book are intended to assist project owners and their agents when embarking on the procurement of contracted services. When followed, they lessen the risk of commercial failure during the phases of contract planning, formation, and administration. Contract controls allow improved contract planning—planning that considers commercial risk and allows owners to select the number, scope of work, and pricing structures for contracts that best meet company objectives.

Formation controls help owners award contracts to capable contractors at the lowest possible price. Well-designed contract and bidding documents ensure commercial compliance and reduce the formation effort. Negotiating effort and evaluation costs are also reduced. Contractor scope of work is well defined, decreasing the likelihood of damaging claims, litigation, and costly change orders.

During the contract performance period itself, owners avoid prepayment and overpayment for contract work. Payment leverage over contractors is maintained and helps ensure adequate performance. Change orders are dramatically reduced in number and in impact, and owners are able to institute only those changes they require—with full knowledge of the cost and effect of each in advance. The frequency and cost of claims is lessened, and owner defense of those claims that do occur is reinforced. Backcharge opportunities are surfaced and pursued to the ultimate benefit of the owner.

Informative contract reporting allows owner management to focus corrective action if necessary. Contract audits provide for direct cost savings, in many cases, and long-term indirect savings when operating and system deficiencies are corrected.

Each of these controls is designed to help owners achieve their ultimate objective: the completion of a quality project on time and within budget. When taken together, they represent a sound program of contract management.

## SCOPE OF THIS BOOK

In dealing with any subject, one must give descriptions of activities, project environments, and functions related to contracting in a context consistent with their application. In the case of this volume, several assumptions must be made to place the information presented in a proper context.

First, the contracting processes will be assumed to be private; that is, there will be little if any discussion of government or public contracting processes and their commensurate regulations, procedures, or contracting agencies. Other than the guidelines under which private contracting occurs, such as contract law, industry practices, and commercial expedience, the descriptions and recommendations presented are governed by sound principles of engineering judgment, accepted project management techniques, and business acumen. Reliance on published regulations for particular contract environments is avoided. In their absence, a pragmatic approach is taken to determine "what works best for all concerned."

Associated topics such as contract terminology, or phraseology, case law, arbitration processes, and the like (acceleration, warranties, hold harmless clauses, legal liability, suretyship, and so on) are not the subject of this volume. Instead, we shall deal with operational aspects of construction contracting—how a contract is best achieved and administered.

Of course, contract activities cannot be discussed in a vacuum. They are extremely dependent upon a multitude of interrelated functions and will vary as to content and application with particular project, organizational, and industry settings. For this reason, we will restrict ourselves to large, capital-intensive projects spanning several years and costing several million (or billion) dollars to complete. We will assume that these projects are commercial or industrial in nature, require extensive engineering and design activity, and are constructed for the benefit of an owner who is vitally interested in the project's ultimate cost, schedule, and technical performance.

Forms, documents, and reports that are included are intended for illustration of the concepts described. As they presently exist, they should not be used or adapted for specific applications. On the contrary, each document should be reviewed solely for the value it has in describing operational controls, and not as a suggested format or recommended content.

With this in mind, the guidelines and suggestions presented here are written from an owner's perspective. It is believed that their application will assist all project participants, including contractors, engineering firms, consultants, construction managers, and so on, in achieving a common goal—successful project completion. In rare cases where the objectives of affected parties are conflicting, or adversary, these are discussed. In general, however, the proper conduct of the contracting functions described on the part of the cwner benefits all project participants. The converse is especially true.

# PART 1

## CONTRACT PLANNING

Before the first steps of the contract formation and administration processes are taken for a major project, several critical decisions must be made concerning the content and management of construction contracts to be employed. This contract planning is an integral part of a broader overall project management plan. Where the project plan would address the project schedule, anticipated cost, technical requirements, and responsibility assignment for the various project participants, the contract plan deals with the contracts to be employed to perform the work in accordance with that project plan. The project plan seeks to answer the questions, What will be done, and when will it be done? Contract planning deals with, Who will we hire to do it, and how will we ensure that it is done at the most economical costs, with the least risk to us? Contract planning, then, seeks to outline the methods by which the business of construction will be conducted and controlled.

Elements of a good contract plan include:

*Organizational and Contracting Strategies.* These decisions concern the delegation of work to project participants and are major considerations of the owner's project management philosophy. Depending on the alternatives chosen for a project, the resulting impact on contract functions will vary widely. We will examine the project organizational approaches used for construction from this point of view—how each choice affects owner risk and needed controls over contract formation and administration.

*Contract pricing alternatives.*   There are an infinite number of ways to price and pay for construction work, and each has succeeded in ensuring performance in some environments though failing in others. We will examine the nature of available pricing alternatives in order to understand the features of control (or risk) inherent to each and to discern the need for various formation and administration controls, selectively applied, depending on alternative pricing structures for construction contracts.

*Contract planning and scheduling.*   The number, respective scopes of work, design requirements, and sequencing of formation activities for each contract are choices that must be made during a project's planning phase. We will present key considerations for the selection of each while addressing the interfaces of contract formation and administration functions with overall project scheduling and coordination objectives.

In this section we will explore these three major topics and relate them to the practices of contract formation, administration, and monitoring, described in later chapters.

# CHAPTER 1

## CONSTRUCTION CONTRACTS: ROLES AND RELATIONSHIPS

The construction of large engineered projects requires that many contractors, subcontractors, material suppliers, manufacturers, and others be selected to carry out the functions that traditionally have not been performed by the engineer, architect, or project owner. Whether the project is an electric generating station, pipeline system, chemical plant, or manufacturing facility, these organizations perform various activities, undertaken for consideration (money) from the owner, and in so doing are assigned various duties and responsibilities as well as rights. These elements are commonly reduced to written form and represent a contract between the owner (buyer) and the contractor (seller).

### CONTRACT DOCUMENTS

The contract documents are the script, then, by which the parties involved perform. As such, their importance cannot be overstated. Contract documents (the "contract") will be discussed in greater detail in Chapter 5. For now, however, we must understand that a construction contract is merely a set of criteria, or expectations, that bind the contracting parties. Construction contracts do not normally contain detailed instructions or step-by-step procedures covering all the activities to be performed. In almost all cases, they are not perfect, exact, or ideal. They are written and interpreted (or ignored) by human beings and applied to complex, high-stakes environments subject to the inadequacies of organizations and the whims of events that are, at times, out of their control, such as weather, strikes, natural disasters, and manpower and resource constraints.

The inherent difficulties of large project construction cause us to seek to control that which is controllable and provide for risks associated with those events or instances beyond our control. This is the case with the functions of contract formation and administration.

## FORMATION VERSUS ADMINISTRATION

*Contract formation* and *administration* are terms representing the business management of contracts. In this analysis, these terms will mean the management of construction contracts—that is, contracts under which construction site labor is expended as a part of the scope of performance. In this sense we shall include contracts where the performer (contractor) provides material, equipment, and labor (furnishing and installing process piping, for example) or only labor and its tools (such as under a contract for foundation excavation).

These twin functions embody activities that are generally divided into two chronological categories: those that are performed up to and including the time a construction contract is established and the contract documents themselves are signed or executed by the contracting parties — called contract formation—and those activities that continue throughout the performance period, ending, it is hoped, with successful completion of the contracted work, final payment, and closeout—called contract administration. Put simply, contract formation is the process of securing a written and signed contract, whereas administration is the handling of the contract until it expires, through natural causes or otherwise.

## THE CONTRACT ADMINISTRATOR

In order to avoid semantics problems and organizational slants, we will assume that one person, or organization, represents the owner in seeing that contract formation and administration activities are performed. For lack of a more accurate title, we will use the generally accepted term *contract administrator* in referring to such a person or organization. As we will see, contract formation and administration are eclectic functions, interfacing with many disciplines and traversing many internal and external organizational boundaries. As such, the contract administrator must possess an eclectic collection of project-related skills and be acutely aware of the needs, methods, and products of other project participants.

As large, capital-intensive projects become commonplace around the globe, spanning many political, organizational, and management orientations, a host of relatively new construction-related specializations, such as contract administration, have emerged to cope with these technical and commercial complexities. Twenty years ago professionals claiming to be cost engineers, risk managers, quality assurance specialists, construction specifiers, or contract administrators would have found it difficult competing for recognition among organizations engaged in construction projects. Today, though the terms may not sound quite so foreign, persons occupied in each of these practice areas are still struggling to establish their identities and to clarify the confusion and misunderstanding surrounding their functions and their project contribution. For the contract administrator, this has been a difficult task for two reasons, described in the following paragraphs.

1.   Contract administration is a relatively new specialization and as such must impinge on real or imagined professional territory collectively occupied by other, more established and recognized participants on the construction project management team. Jurisdictional disputes involving authority, responsibility, and interface relationships are common. This condition is particularly aggravated by the high degree of visibility contract administration enjoys among financial and legal management—visibility that technical personnel sometimes envy or suspect. The simultaneous emergence of other equally novel professions, discussed earlier, further complicates the establishment of contract administration as a distinct and viable occupation.

2.   There is no generally accepted definition of contract administration or of the qualities and skills required of a contract administrator operating in the private sector. Unlike the legal and engineering professions, where accredited universities dispense training and diplomas, governmental agencies examine, license, and regulate practitioners, and professional societies establish ethics and codes of behavior for their members, contract administrators spring up from everywhere and are guided or sanctioned only by their own wit and the demands of the marketplace. One can safely predict that, given sufficient time, associations and other social structures will surface to control the quality of those holding themselves out to be contract administrators and to prescribe the boundaries and conduct of their practice. As of today, only limited and local attempts at such have been made. As a prerequisite to more general or universal regulation and guidance of contract administration as an occupation or profession, we must first achieve a common understanding as to what, in fact, it is.

## PERFORMANCE MONITORING VERSUS CONTRACT ADMINISTRATION

In order to operate at arm's length with a construction contractor and to maintain the independent contractor relationship, the project owner and its representatives must refrain from exercising direct control over the means and methods employed to construct the work. These are left for the contractor to control. Although the contractor will provide the labor, material, and equipment necessary to fulfill the commitment, it is also solely charged with supervising that the work is performed adequately.

The contractor, therefore, must perform and has the right to control the work. The owner, on the other hand, has the right to expect that the project is being designed and built according to its requirements and established standards of quality. In this capacity, the owner does not supervise but rather gives the contractor a set of expectations to meet, inspects and tests the work as it is developing (and when completed), and either accepts or rejects the results. *Performance monitoring* is a term used to describe this more passive, less direct involvement in the activities of the contractor. Specific techniques employed by owners to assure themselves that contractors are producing work in accordance with the contract requirements include:

Inspection

Testing

Review and/or approval of contractor submittals

Guarantees, warranties

Independent certification

Quality assurance criteria

Acceptance requirements

These are all elements of performance monitoring, not of contract administration, and the distinction is significant.

Whereas performance monitoring and its associated corrective action answer the question, Are we getting what we are paying for? contract administration addresses a variety of ancillary questions. These range from, Is the contractor maintaining adequate insurance coverage? Are changes in the work being priced correctly and paid for expeditiously? and, Are proper costs records being maintained? to Are claims or disputes being handled on a timely and fair basis?

**FIGURE 2.** Contract responsibilities.

Figure 2 summarizes the contract responsibilities of the owner's representatives and the contractor.

## Conflict of Interest

Although both performance monitoring and contract administration may be undertaken simultaneously by the same individual or group of individuals, each is a distinct function requiring special talents and training. Even though they are closely associated by their very nature, a conflict of interest may develop when performance monitoring and contract administration are handled by the same person(s). For example, a construction engineer representing the owner may find it difficult to see that a certain section of work be completed by a contractor in order to meet schedule commitments, even though the work is not within the scope of the existing contract. In his or her zeal to see this accomplished, the engineer may be inclined to forgo the necessary contractual modifications (including cost quotations, acceptance by the owner, and written orders to proceed) that are contract administration functions.

Traditionally, the duties of formation and administration of construction contracts have been assumed by the project owner or, in many cases, the party responsible for preparation of the contract documents—such as the architect or engineer. As the one most intimately associated with the intent of the contract documents, especially when it has prepared them and will oversee their application, the designing organization might seem best suited to perform the duties of contract administration. Care must be taken to avoid assigning the responsibilities of performance monitoring and contract administration to the same individual, or to allow a functional organization responsible for one to dominate the other.

On large projects with many prime contractors, all working under separate agreements with the owner (sometimes as many as 100), a case can be made for centralization of the administration process, leaving the performance monitoring to several individuals well-versed in the particular trade or construction process they are monitoring. Just as uniformity among all of the separate construction contract documents is a primary goal of those preparing the documents, so should uniformity and consistency be attempted in the administration of such contracts. Allocating all contract administration duties to one individual or group of individuals, especially on multiple – prime contract jobs, is highly desirable.

## GENERAL RECOMMENDATIONS

A review of the functional topics contained in this volume should give an indication of both the duties and identity of a professional contract administrator as well as proper formation and administration practices. Each separate function reinforces the principle that successful contract administration depends upon a high level of communication among the contractor, contract administration, and owner.

Emphasis is also placed upon proper documentation of all aspects of the administrator's activities. The administrator must be meticulous in record keeping, as well as handling of working correspondence and documents. Likewise, he or she must expect rigid adherence to provisions of the contract yet be flexible and open-minded enough to handle unforeseen conditions or unanticipated changes, which are inherent to the construction process.

The ability to get along with people from many social and economic backgrounds and representing many diverse interests also is essential in the character of a contract administrator. Coupled with a sizable amount of tact, it will preserve both his ability to perform and sanity.

### GROUND RULES

Once a contract administrator has been chosen, his or her first mission is to establish ground rules covering conduct and relations with others during the period of the assignment. Failure to define and reach agreement on his "scope of work" would plague him throughout the duration of the project, just as much as would failure to agree upon a contractor's scope of work. Basic ground rules should cover the areas of responsibility assigned to the contract administrator (and—just as importantly— those *not* assigned to him), working and reporting relationships vis-à-vis other members of the organization (project management, the construction manager, estimator, accountant, engineers, and so on), and his authority (or limits of authority) to initiate or approve actions in the interest of the owner.

As important as a clear description of duties is to success, it is imperative that the administrator have a thorough knowledge of all contracts he is to administer. He must establish himself as the one expert when it comes to the content of the contract documents. Often he is called upon to defend the owner's position in a dispute by relying on their provisions. To refer to what he "assumes" is in the contract only to find that the contract states otherwise could be particularly embarrassing, especially when brought to his attention by a contractor. Besides suffering embarrassment, the administrator could easily destroy credibility he had worked so hard to establish and without which he could not operate.

Many times the administrator will have to modify a contract to incorporate extra work or changing conditions or requirements. He or she could not enter into the documents to amend them without knowing what he is amending. He should not change any part of a contract without a complete awareness of each provision he is changing and of the impact of the change(s) upon the document as a whole. A common mistake is to assume that the contract documents need changing to provide for an unexpected or new event, only to discover that the existing documents adequately cover such an instance. This is not to imply that the documents are somehow sacred and not to be tampered with, but merely to make the contract administrator aware that changes must be made with caution and with a knowledge of the need for and impact of such changes.

One final point in this regard must be addressed. Contract administrators, due to the very nature of their work, can easily forget or overlook the need for review and advice from qualified legal counsel. It is usually impractical to insist that contract administration be performed only by attorneys, and it is paramount that those who are not qualified to practice law refrain from doing so. On the other hand, the sole reliance on legal counsel to prepare technically and commercially sophisticated contract documents is just as grievous an error. Lawyers may be needed to review, create, or comment on the contract documents as a whole and specific clauses in particular, but they are generally unfamiliar or unconcerned with the operational ramifications of contract wording. Legal review usually limits itself to questions of liability and enforceability. Project management and contract administration, by virtue of their experience, education, and vested interest in the day-to-day problems that contract wording, or omissions, may create, have a different perspective. For example, although a lawyer may point out the proper "hold harmless" or insurance requirements needed, he or she should not be relied upon to ensure that the contract ties progress payments to specific, quantifiable milestones or that as-built drawings be approved as a prerequisite to final payment. Once again, as is so often the case, common sense and good judgment should prevail.

## A SYSTEM IS ESSENTIAL

Without a logical system by which to operate, the contract administrator will be lost in no time. His or her job is to bring order to the contracting process. He cannot do this job without order in his own activities. A true contract administrator takes an active role in the daily operations of the contracts. Without adhering to a minimum amount of regimentation, he can fast become a passive member of the construction team reacting to events rather than foreseeing or controlling them.

A common way of establishing a system is through project procedures and working instructions, such as those contained in this book. This up-front planning will save countless subsequent arguments concerning who does what to whom. A good set of working tools must be built before the administrator assumes the assignment. A filing system should be established, as should be the format and forms of paperwork he will handle. Simplicity and uniformity should rule when developing standard documents for common transactions such as contract changes, back-charges, progress estimates, and contract closeout. Checklists, logs, and standard forms, such as those included in this book as exhibits, are extremely helpful in bringing order to operations. Not only will they simplify work by streamlining it as much as possible, but they will also provide the basis for the airtight documentation required for claims management and for concise and consistent record keeping—a blessing when audit time arrives.

All of the above should not be construed as making the contract administrator a glorified clerk. This would be a degradation of the function both in theory and in practice. It is indisputable, however, that a certain amount of "everything in its place and a place for everything" thinking is not only tolerable but also desirable. Should the contract administrator develop a crisis-oriented mentality—responding to events rather than controlling them—he would drown quickly in a sea of disorganized paper.

## THE REQUIRED SKILLS

As this book indicates, contract administration serves as a staff or advisory role rather than as an organizational line position. The contract administrator does not enforce contract provisions, nor does he or she manage the contractors or the project. These are the responsibilities of the project manager. The contract administrator represents the owner or project manager as a business advisor regarding the commercial and operational aspects of the contracts for construction.

As is often the case in the industry, the very skills required for contract administration are not generally offered in our technical schools. Of course, technical background, either from engineering college or gained through experience, is necessary, for the contract administrator must have a concept of what is being constructed and the significance of design and product specifications. Common business sense, gained through formal training (business administration, contract law, accounting), or through experience, also is required.

An ideal background includes familiarity with construction operations and methods. This encompasses not only the construction process itself but also the environment in which the industry operates. General knowledge of construction financing, insurance, equipment, labor relations, and supervision are some topics in this category. Although there are not complete contract administration curricula available—as there are for other occupations—excellent short courses are beginning to appear through various organizations associated with the construction industry.

## UTILIZATION

A lot can also be said concerning how and when the administrator is utilized. Ideally, he or she has had a hand in the preparation of contract documents and the formation cycle and follows them to the job site to administer them throughout the construction period.

As much as possible, the functions of contract formation and contract administration should not be performed by separate entities within the owner or design professional's organization. Experience in both major aspects of contract management should be encouraged, so that

**FIGURE 3.**  Symbiotic learning.

appreciation of one will benefit performance of the other. If this is not practical, and home office preparation and award must be separated from field administration activities, it is recommended that extensive communication between the two be maintained and that home office support, both technical and administrative, be provided to the person administering contracts in the field office. This symbiotic relationship between the two major phases of contracting is illustrated in Figure 3.

## A DISTINCT FUNCTION

Contract administration is a distinct function best performed by individuals trained in a variety of fields and endowed with certain character traits. Although contract administrators have suffered because of a lack of identity, a certain amount of knowledge exists to establish them as professionals in the construction industry and as important members of the construction team. Basic guidelines can be established for the training, selection, and utilization of contract administration personnel as they emerge from obscurity in today's complex construction contracting environment.

# CHAPTER 2

# ORGANIZATIONAL AND CONTRACTING STRATEGIES

Once an owner decides to embark on a construction project, the major decision it must make is how best to organize the effort and delegate work to outsiders. This decision affects not only the management of the project but also the number, scope of work, and responsibilities assigned to internal and external organizations involved. This chapter describes major organizational alternatives, considerations for their selection, and the resulting impacts such a selection has on contracting processes.

## CONTRACTUAL IMPACTS OF THE ORGANIZATIONAL APPROACH

Because the purpose of this chapter is to indicate the impact an owner's approach will have on subsequent contract management activities, rather than the merits of each approach alone, it is important to understand the far-reaching ramifications of such a decision. First of all, the chosen organizational approach determines the number of contractors that will be involved. In this manner, it defines the extent and limits of contracting — dictating the contract formation and administration efforts required.

The organizational approach used also sets the framework for project-related authority and general responsibility assignments. The construction contracts further expand and define this authority and responsibility among the organizations involved. They follow the pattern established when an organizational approach is selected by assigning roles and responsibilities to those who will be operating under contractual relationships. In addition, because the selection of an organizational approach implies a general scope of work for each participant, the construction contracts musts clarify and communicate the details of that scope in order for the owner's ultimate objectives to be met. Construction contracts, then, may be viewed as the tools by which an owner implements the specifics of the chosen organizational strategy.

## ORGANIZATIONAL IMPACTS OF THE CONSTRUCTION CONTRACTS

Not only does the project's organizational approach affect the resulting contracts used to implement such a strategy, but the converse also is true. That is, the contracts themselves impact the organizational approach. For example, the contracts establish project-related authority, they distribute scope of work among project participants, and they specify the performance standards that must be met by those under contract with the owner. They also define the consideration—that is, money — each participant is to receive for project-specific work and identify the technical and commercial controls for the contract and the project as a whole.

When discussing contractual controls, therefore, we must realize the close association between the contracts themselves, the various contractual relationships they establish and enforce, and the project organizational setting in which they apply.

## CONSTRUCTION'S PROJECT ORIENTATION

The construction of a major facility differs markedly from other large efforts requiring a great deal of time, money, and human resources. The most important difference is that construction is "project oriented"— that is, a unique collection of people, equipment, and material is brought together to create a finished product at a unique location. Although most of the tasks required to construct a facility have been performed many times before, the collection of tasks performed by the particular people, at a particular time, at a unique location, and for the particular needs of an owner constitute a one-time, unique project.

Contrasted with line consumer manufacturing—the mass production of automobiles, for example—where a relatively constant set of people, equipment, and activities produce a relatively constant product over and over again, construction of a facility occurs only once. In most cases, the collection of resources—people, material, equipment, and so on—needed for a project has never been brought together before and will never be assembled in the same manner for the same purpose again. The only common point of reference for all project participants is, therefore, the project itself.

If one looks at the example of automobile production from an organizational perspective, it is fair to assume that an operating plant producing these items does so under an established, ongoing organizational structure. This structure exists under the authority of one company and uses established lines of reporting and communication. Each member of that organization, presumably, understands his or her role and performs activities in order to meet individual long-range goals and those of the organization.

These features are not evident with a construction project. First of all, the organization involved is not established and operational. It must be created. Lines of communication need to be established, and there will be many authorities involved—sometimes operating with conflicting priorities. Because the project represents a temporary, one-time effort, short-term goals and relationships predominate over the long-term variety.

Functional contrasts may be drawn between the automobile production example and construction's project orientation. Line manufacturing requires the repeated production of identical or similar products using a series of tasks performed over and over again. Construction, on the other hand, involves unique, one-of-a-kind products, produced through a unique set of tasks. Production steps taken with mass assembly are linear—that is, they are performed in a structured sequence. This is not the case with construction. It involves steps that take place collaterally, or in parallel. Transient production facilities and methods are used in construction, such as mobile cranes, specialized rigging arrangements, and portable welding machines, whereas line manufacturing's facilities are fixed.

A primary difference involves the ultimate owner of the finished product in the production process itself. For line manufacturing, the owners (in this case, the potential buyers) have no direct impact on the manufacture of the product—they are not even identified until the products are completed. Owners engaged in construction efforts, on the other hand, are vitally interested in each step of the effort as it progresses. They choose the project participants, delegate responsibility to each, and manage or monitor the project throughout its life.

Many of these distinctions may seem fairly self-evident, but each affects every aspect of the construction industry and has shaped the way its business is conducted. Each points out the importance of the contracts themselves, as well as the demands of contract management. For it is the contracts that establish the roles and relationships that must be created, define the production organization, and provide incentives and penalties in lieu of long-term relationships and goals that exist elsewhere. Considering the project-oriented nature of construction, then, the contracts under which it is performed and the management of those contracts represent major factors for success or failure.

## TYPICAL ORGANIZATIONAL AND CONTRACTING STRATEGIES

There are a myriad of ways by which an owner may organize and manage a construction project. Each has proven successful when properly planned and implemented. And each has established the framework for failure when selected by owners unfamiliar with its features, strengths, and concurrent owner obligations—or when inadequately implemented. In describing the features, benefits, and disadvantages of each, two important facts must be kept in consideration:

1.  No approach is inherently superior in all cases. Each approach described has been used favorably and unfavorably in a number of cases. Although proponents of various approaches have attempted to justify the superiority of each, there remains no consensus in the industry that would lead one to choose any approach over the other without considering other owner- or project-specific factors.
2.  Combinations of approaches have been used and are fairly common. There is no reason for owners to ignore possible "hybrid" approaches incorporating aspects of several described here. In fact, many owners have employed different aspects of each approach during different phases of a project's life cycle, or for different portions of the project. Indeed, it is seldom the case that one "pure" organizational approach is chosen and maintained throughout a project's life, particularly when the project extends over several years. Quite often, variations are adapted to meet specific or changing owner objectives, economic conditions, and management philosophies.

Five general approaches are described in the following sections. They are represented by generic organization charts in Figure 4 in an attempt to point out salient features and differences among them. In addition, an example "hybrid" approach is also illustrated to indicate only one of many thousands of combinations available.

**FIGURE 4.** Organizational approaches.

31

## DESIGN–BUILD

This is an extreme example of the project owner delegating responsibility to an outside party. As we shall see later, it is the opposite of the final approach depicted (force account), where the owner assumes the dominant, if not total, responsibility for the project's completion.

With the design–build approach (also called a turnkey contract) the owner selects one firm to both design and construct the facility. It gives the design–build contractor a set of criteria it expects the completed project to meet, such as operating and performance characteristics, cost, site location, aesthetic appeal, date required, and general preferences—for example, similarity to an existing facility, use of local materials, architectural style, and brands of equipment preferred. Within these general parameters, the contractor must design the project and build it as well. In some cases, the contractor agrees to operate the facility over a period of time, either selling the products to the owner at a predetermined price or leasing the facility from the owner at a prescribed fee.

Oftentimes the design–build contractor chooses to subcontract portions of the work to others. The design–build concept, however, still remains because the prime contract between the contractor and the owner— not the degree of subcontracting—is the distinguishing characteristic of this approach.

The assignment of responsibility under a design–build approach is straightforward. Except for certain owner duties, such as payment for work performed, review and approval of intermediate and final products, and general decision-making obligations, the majority of project responsibility lies with the design–build contractor, who is responsible for planning, conducting, and coordinating all tasks necessary to produce the finished facility. Even with this concept, however, there are many responsibilities that are assumed by the owner unless specifically assigned to the contractor or others by the contract. Such activities and responsibilities must be carefully considered and addressed by the owner prior to engaging a design–build contractor. Among these are licensing, real estate acquisition, permits, design approval, accounting and financial control, reporting to outside agencies, safety, and security.

A major commercial risk associated with the design–build concept concerns the lack of management flexibility it affords the project owner. By the selection of only one company both to design and to construct the facility, the project's success becomes directly dependent upon the financial stability, management controls, and operational effectiveness of that company. Should any prove unsatisfactory, it is difficult to remove the contractor from the project without severe impact on its cost, schedule, and technical performance. These impacts become greater as the project progresses. As will be discussed further in the following chapter, the design–build approach is rarely contracted in a firm-fixed-price manner. Because of the inherent uncertainty in scope of work (that is, the design has yet to commence) cost-reimbursable pricing structures are most often employed when contracting for design–build services.

From an owner's standpoint, the most critical performance control is the selection of the design–build contractor. Since detailed controls over cost, schedule, and technical performance will be implemented by the contractor, the owner must satisfy itself that these controls are adequate before the contract is signed. This requires extensive bidder qualification and a detailed examination of specific contractor controls during the selection process. Requirements for contractor controls should be specified in detail in the contract documents, and the owner must monitor the use of these controls during execution of the project. Because the owner, by nature of the approach, must delegate detailed control responsibility to the design–build contractor, it must assure itself that the contractor has the intent, personnel, and systems to exercise such control. The owner's duty then reverts to surveillance of the contractor's controls throughout the project duration.

As we shall see later, the degree to which the construction project must be defined or specified prior to awarding a design–build contract is slight. This is an advantage to its use. Additionally, there is only one contract to form and administer from the owner's point of view. Impact on the owner's staffing and management requirement is minimized with this approach.

## GENERAL CONTRACTOR

This is the traditional contractual approach most often seen in smaller building construction projects, such as hospitals, churches, schools, and shopping centers, but used quite frequently for industrial projects as well. With this approach the owner contracts with an outside engineering and design firm (architectural–engineering firm, or A–E), which will design the facility, and also contracts separately with a large construction contractor to build it. Most large general contractors will, in turn, subcontract portions of the construction work to other smaller or more specialized contractors (subcontractors, or "subs").

The difference between this approach and design–build is that the design function is awarded separately from the build function. A third party, the A–E, is brought under contract with the owner rather than as a part of or subcontract to the constructor's organization. The owner is now dealing with two distinct organizations under contract. Both— A–E and general contractor—must interface with each other excessively in this arrangement. As opposed to the design–build situation, the awarding of the design contract precedes award of a contract for construction. The products of the design effort (drawings, specifications, and so on) are needed to allow general contractors to determine the scope, cost, and schedule of their work. Ideally, all of the design is completed prior to the bidding of a construction contract. The work is well defined, a schedule can be readily determined, and a price quotation can be made by the bidding contractors with some degree of confidence. In actuality, this is seldom the case.

The general contractor approach provides for a classic division of project responsibility between engineering and construction efforts. The owner is not typically involved in subsequent delegation of responsibility within either of these contracted efforts. And coordination of work performed by any subcontractors to either the A–E or general contractor is the responsibility of these respective companies.

Two major challenges confront owners contemplating this approach:

1.  Identification and assignment of responsibility for those functions that, though required for all projects, are not automatically associated with either design or construction. This includes such functions as licensing, permits, procurement, and vendor expediting, among others.
2.  Coordination and communication between the design and construction companies throughout the project life.

Some owners using the general contractor approach delegate much responsibility for non-design-specific functions to the A–E firm. Depending on the extent of this delegation, the A–E may also perform other functions as a representative or agent of the owner. Some functions in this category are procurement and procurement-related activities, project administration, contract administration, and inspection and testing. To the extent that this occurs, the owner may greatly reduce its direct project involvement. In any case, however, the owner must be prepared to coordinate, and sometimes mediate, between these two contractors (design and construction) before selecting this approach.

The "one-contractor" risks inherent in the design–build concept also apply to the general contractor alternative, with the exception that engineering and construction are performed by separate companies.

Since design is rarely performed by more than one company, this effect is more meaningful in the construction context. Reliance on a cost-reimbursable pricing method for the construction portion of the work is not as severe as with the design–build concept. In cases where significant design has been performed before a general construction contract is awarded, the benefits of fixed pricing may be pursued. Two factors contribute to the fact that firm fixed pricing is seldom encountered, however, for the entire scope of construction:

1. Design is seldom totally or substantially complete at time of construction contract award.
2. Inflation, escalation of material and labor prices, and changes in the work invariably occur during the length of time required to perform all project construction work. These changes, as we shall see in the next chapter, are normally reflected in unit price or cost-reimbursable pricing clauses in the contract.

The added burden of coordination placed on the owner also increases demands on owner staffing and system requirements. To the extent that owners do not meet the challenge of this increased involvement, the risk of unsatisfactory performance increases.

Because the coordination of these two major efforts is typically the responsibility of the owner, owner controls must be put in place as soon as an A–E is selected. These include provisions for sequencing of design work to meet the needs of construction, timely and adequate review of design products (drawings and specifications), schedule management, and communication control. When owners accept direct responsibility for functions that span design and construction, such as procurement, scheduling, and project reporting, owner controls over these functions must be implemented.

## FEW PRIMES

The few-primes approach is not as distinct as others described in this chapter. It can best be thought of as representing the "middle ground" between the extremes of design–build and force account. It occurs when the owner awards more contracts for construction than are used for the general contractor approach—that is, more than one—yet fewer than those encountered by owners under a multiple-prime strategy. A typical few-primes situation is where separate prime contracts (as opposed to subcontracts) are awarded between the owner and, say, five contractors. In most cases the subdivision of work among these prime contractors is along product or discipline lines, such as civil, mechanical, and electrical. There may be one contractor designated as the "general," but all other contractors are not necessarily subcontracted to it.

For example, the owner may award separate contracts for inspection and testing services, painting, and electrical work, with everything else being performed by the general contractor or subcontracted by him. Indeed, the owner may also elect to buy certain equipment and material from suppliers and furnish these to contractors for installation during the construction process as well. The degree to which the owner becomes directly involved in the selection of contractors and suppliers determines the associated responsibility it must assume for their coordination. Under the umbrella term of *few primes*, this involvement on the part of the owner may vary greatly.

As with previously described alternatives, the design function is performed internally or given to one outside firm. The project construction effort must be thoroughly planned, packaged into distinct scopes of work, and awarded in such a manner that each item of work is assigned to one and only one prime contractor. This effort becomes more demanding and risky as the degree of contracting employed increases and the allocation of work scope becomes less traditional.

Under this approach the owner assumes many detailed coordination, administration, and scheduling responsibilities normally performed by the general contractor under the approach using that term. The owner must insert itself not only among the few construction contractors but also between each contractor and the A–E firm. In some cases owners engage the services of an outside "project management" or "construction management" firm to discharge these coordination duties. This is not a requirement of the few-prime approach, but it often eases the detailed demands on the owner's staff and systems.

The risk of investing all construction work with one contractor is reduced with the few-primes approach to the extent that more contractors are involved with the construction effort. Should any one contractor experience stability or performance problems, it may be more easily terminated or replaced than if it were responsible for all construction work.

Cost risk may be reduced through the selective use of firm-fixed-price contracts, sequenced for award after complete design information is available. Shorter construction durations for each contractor also reduce reliance on cost-reimbursable pricing. The greatest element of risk to a project owner operating with this approach involves the use of smaller (perhaps less stable) contractors whose performance is greatly dependent upon owner management, coordination, and control capabilities. The lack of a general contractor to schedule, sequence, and coordinate the construction work, including the administration of construction disputes, requires the owner to increase its project involvement. To the extent that an owner fails to fulfill these obligations, its risks increase dramatically.

As with the multiple-prime approach, basic control methodology on the part of the owner undertaking this approach centers on:

1. Discrete, well-defined subdivision of work among the associated contractors
2. Thorough qualification of potential contractors
3. Selective control of cost, schedule, and technical aspects of each contract based on performance risks
4. Proper level of owner involvement to manage and control each contractor's work

## MULTIPLE PRIMES

When there is no dominant or "general" contractor, and most of the organizations associated with the construction are performing under separate, direct contracts with the owner, a multiple-prime situation exists. In a sense, the owner has replaced the general contractor in its role of subdividing the work and awarding individual subcontracts (in this case, *contracts*). Needless to say, this increases the owner's participation one level greater than the previous approach. The owner (or its agent) becomes the ultimate coordinator of all contractors on the site. Often, the multiple-prime approach is called *construction management*—a term that is as difficult to define as is contract administration. It also requires the issuance and administration of more contracts than do all other approaches. The main disadvantages of a multiple-prime approach are easily apparent: extreme demands on the owner (and A–E) for staffing, responsiveness, and organizational strength. Many owners fail to realize this and undertake this approach without preparing for the increased responsibility, additional staffing, and assumption of risk that it entails.

The multiple-prime approach does have its advantages though. For one, the packaging of work into many small contracts makes the use of smaller local contractors possible. In addition, a type of "portfolio effect" is achieved. The owner is not captive or unduly dependent on one large contractor once construction is under way. It has the flexibility of awarding each package on its own merits as far as pricing structure is concerned. Should, for example, one segment of the work be undefined, such as landscaping, the owner can hold off on its award until it is defined. Once construction begins and several contractors are mobilized, the possibility of their being awarded later portions of the work, should they perform well, is a valuable incentive.

A condition called "fast tracking" (either planned or unplanned) exists when portions of the design, or construction "packages" are released from the A–E to the contractor(s) in advance of other design products, which are in preliminary stages of development and are not immediately required. This is advantageous in getting the construction started as early as possible—for example, foundations can be excavated and poured before electrical equipment and wiring has been designed—but it intensifies the scheduling and coordination demands on the owner, A–E, and construction contractors.

Because the multiple-prime approach lends itself to the staggering of small design and construction efforts, it is commonly seen when fast tracking is used. Fast tracking has been heralded as a viable time-saving technique, with overall project schedule compression as its advantage. It requires that design be several steps ahead of construction. However, at times design falls behind or rushes out incomplete and unsatisfactory products. Therefore, rework and changes to the construction contracts are common.

## FORCE ACCOUNT

Force account is the owner's ultimate step toward project involvement and responsibility. It presents the classic "do it yourself" approach. In its extreme case, the owner designs and/or builds the entire project, using its own personnel and equipment exclusively. As in all approaches described here, the owner has dedicated personnel assigned to the project. But with the others, the owner's people perform management, supervisory, or monitoring functions. With force account, the owner also has the actual craft workers (carpenters, ironworkers, bricklayers) on its payroll. The workers (forces) are to its "account."

Usually, even with a dedicated force account owner, some specialty work is contracted to outsiders. The Tennessee Valley Authority, with its electric generation station construction program, is an example of a very large owner that uses the force account approach. Even with the TVA, some special work—furnishing and erecting nuclear plant containment liners, for example—is awarded to outside contractors.

A variation of the force account approach occurs when, rather than placing construction craft workers on its payroll, the owner hires a "labor broker" to furnish this manpower. With the labor broker method, the broker differs from a contractor in that it assumes no risk for contract completion and is not responsible for the work itself. This is left to the owner, with the labor broker simply furnishing manpower on an as-required basis. For this service, the labor broker is paid a fee, which is generally a percentage of the total wages paid to the workers he provides. The workers are employed by the broker but used and directed by the owner.

Needless to say, a force account approach places extreme demands on the part of the owner. For this reason, most owners—except for those with extensive long-term construction programs—avoid this strategy. Even continual builders, however, cite the following reasons for passing up the opportunity to go into the construction business:

1. Possible adverse reaction of outsiders (contractor organizations, stockholders, and so on)
2. Limitations on available manpower
3. Personnel recruiting, training, and retention costs
4. Need for large equipment and supply inventories
5. Construction labor relations difficulties
6. Increased liability for construction-related tasks such as transportation, logistics, security, and safety

## SUMMARY OF MAJOR FEATURES

In reviewing the five major organizational approaches described in this chapter, several conclusions may be drawn. These are summarized in Figure 5 and described in the following paragraphs. Only by thoroughly understanding the features and benefits of each approach can owners: (1) make a rational selection of any one approach over the others; and (2) successfully implement the choice they have made.

1. *Level of Owner Involvement Required.* The major effect of an organizational approach is to dictate commensurate owner involvement in project-specific activities. And depending upon the subsequent assignment of responsibilities within the boundaries of any given approach, the owner's involvement varies greatly.

2. *Contract Pricing Methods.* Although described in more detail in the next chapter, pricing alternatives are dependent to a great extent upon the organizational setting of a construction project. Absent other influencing factors, such as the competitive nature of the contractor

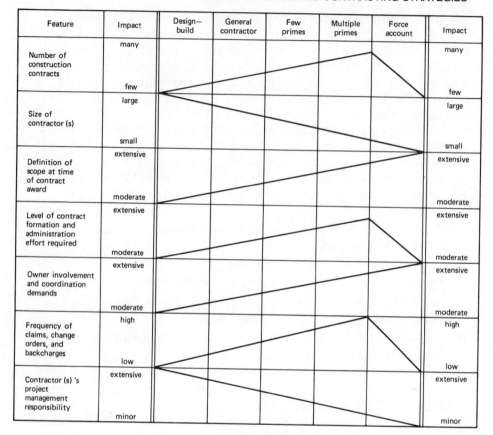

**FIGURE 5.** Impacts of contracting approaches.

marketplace, firmness of design information, and general economic uncertainty, the ability to secure firm-fixed-price contracts increases as each contractual performance period and scope of work decreases.

3. *Single Contractor Risk.* As the number of outside companies involved in a project increases, the vesting of risk in any one decreases. Negative impacts resulting from failure, termination, or inadequate performance by any one contractor are reduced when the total project scope is distributed among many outside organizations. The benefits of this "portfolio effect" on risk must be weighed against increased management and control demands.

4. *Contractor Control Capabilities.* As the relative scope of work and size of a company increases, its internal management and control

capabilities typically improve. In this regard, larger design–build contractors usually maintain more extensive management and project control capabilities in comparison to smaller, specialized contractors who typically work on restricted scope, within the environment of controls provided by others.

5. *Contract Formation and Administration Requirements.* Two major factors influence the scope of contract management duties and the difficulty with which they are performed. These are the organizational approach selected by the owner and the pricing structures used for each contract. Absent pricing considerations, contract formation and administration efforts increase in direct proportion to the number of contracts used for a given project.

# CHAPTER 3

# CONTRACT PRICING ALTERNATIVES

There are an infinite number of ways to price and pay for contracted goods and services. In this chapter we will explore the most commonly used methods from the viewpoint of the features, element of cost risk to the owner, owner control methodology, required level of owner involvement, and applicability of each.

The most common ways to price contracts are shown in Figure 6. They are arranged in major groupings of those considered fixed-price techniques and the remainder in the cost-reimbursable category. As shown, they range from (1) firm fixed price or lump sum, where virtually all cost risk lies with the contractor, to (2) Cost plus percentage of cost, where all the cost risk rests with the owner and, in fact, the contractor has a positive incentive to increase the cost of his work. Within these extremes, there are almost unlimited possibilities and combinations. Because of the range of these alternatives, it is sometimes impossible to categorize clearly a given contract as either "fixed price" or "cost reimbursable," as it may encompass aspects of both in an attempt to incorporate reasonable incentives and risks from the standpoint of both the contractor and owner.

## PRICING TERMS

In our discussion of various pricing alternatives, several terms will be used. Each has a precise definition, and only by understanding the

**FIGURE 6.** Typical contract pricing alternatives. (Pricing methods marked * are seldom encountered in conjunction with the construction of major commercial or industrial projects.)

meaning of each can one expect to understand, or unravel, the sometimes confusing distinctions encountered with pricing alternatives used in the construction industry today. Basic terms used with regard to each alternative are:

1. *Cost.* This is defined as the seller's (contractor's) actual cost; that is, the price it pays for labor, equipment, material, and so on.

2. *Fee.* This represents an additional amount to be paid to the contractor over and above its incurred costs. It usually, but does not always, represents payment for his overhead, profit, and expenses not included in the above-defined cost. As will be seen in the descriptions of some pricing alternatives, this fee may be positive (contractor gain) or negative (contractor loss).

3. *Price.* Price represents the total amount paid by the owner to the contractor. In most cases, it represents the sum of the contractor's cost and fee. The contract price can also be described as the owner's "cost" for having the contracted work performed.

## CONSTRUCTION CONTRACTS VERSUS PURCHASES

As with all discussions contained in this volume, the term *contract* refers to those contracts in which job-site labor is involved in the seller's scope of performance. This distinguishes *contracts* from *purchases*, where no job-site labor is involved on the part of the equipment or materials vendor. An example of a purchase would be the furnishing of pumps, motors, or electrical cable to the owner or to a contractor by a supplier. Of course, job-site labor would be required to install these items, and the installation would be covered by a separate construction contract of some kind.

As mentioned earlier, construction contracts can include the furnishing of labor and its tools alone, such as foundation excavation, or labor and materials or equipment. These latter contracts are called furnish-and-erect or furnish-and-install contracts, and an example is a contract for furnishing and erection of structural steel. The pricing alternatives described in this chapter can be used on conjunction with either furnish-and-erect or labor-only contracts.

With the above distinction in mind, most owners prefer the use of different contracting instruments, depending on whether they are buying construction contracts or purchases. Construction contracts are used to establish the former, whereas purchase orders are typically used for the latter. The distinction between these two instruments is further developed in later chapters.

## CONSIDERATIONS FOR SELECTION OF PRICING METHODS

Several factors influence the advisability of choosing pricing alternatives in any given situation. They affect the contractors' ability to provide an accurate and firm estimate of the cost or price of performance as well as the ease with which progress can be measured and paid for during the performance period. Each of the considerations described below should be made by owners before deciding upon contractual pricing methods they may wish to use.

1. *No single pricing method is superior in all cases.* Although a general consensus among owners in the construction industry favors fixed-price types of contracts over the cost-reimbursable variety, the latter may prove far more cost-effective given certain conditions. Each pricing method may be totally appropriate given certain owner objectives, scope of work, bidding information and bid circumstances, and market conditions and yet wholly inappropriate given others. The selection of

a proper contracting strategy (including pricing method) should be based on a thorough analysis of these factors on a case-by-case basis. In general, every effort should be made on the part of an owner to plan and control contracting efforts to facilitate fixed-price buying. However, it must be recognized that cost-reimbursable contracting does have benefits when appropriately used, carefully defined, and closely monitored. General relationships among pricing methods and both owner cost risk and cost monitoring required are shown in Figure 7.

2. *Combinations of pricing methods are common.* It is not uncommon for owners to use several different pricing methods within the framework of a construction project. In addition, many contracts contain provisions for a combination of pricing methods, each applying to different scopes of work or under differing contractual or project conditions. An example of such an approach would be a contract for soil and rock excavation. A lump-sum (fixed-price) amount may be agreed to for the major foundations, whereas unit prices are employed for additions or deletions (adds and deducts) to the anticipated volumes of soil and rock. Cost-reimbursable terms may also be employed for additional work not included within the scope covered by the lump-sum amount, such as clearing and grading of access roads yet to be designed. If the performance period is long, escalation terms applying to increases in material and labor costs may also be applied. This example illustrates that not only can different pricing methods be employed within the boundaries of a single contract but also in many cases such a combination is more desirable than strict adherence to a single method.

3. *Performance incentives and escalation provisions may be added to each pricing method.* Regardless of the pricing base upon which goods or services are bought, additional pricing criteria are often included in owner–contractor agreements. These include bonuses and/or penalties based upon the cost, timeliness, or quality of the goods or services provided. Escalation provisions, designed to protect the contractor in the event of increased costs due to inflation or other factors beyond its control, are also quite often included. Performance incentives (both positive and negative) and escalation provisions are described in more detail later in this chapter.

4. *Pricing methods must be chosen deliberately, defined properly in contractual agreements, and carefully monitored to achieve their objective.* From an owner's perspective, selection of the best pricing method (or combination of methods) is only the first step in a series of tasks required for cost and performance control. Precise definition of the pricing basis must be included in the request for proposal (RFP) and contract documents. Methods for measuring, accepting, and paying

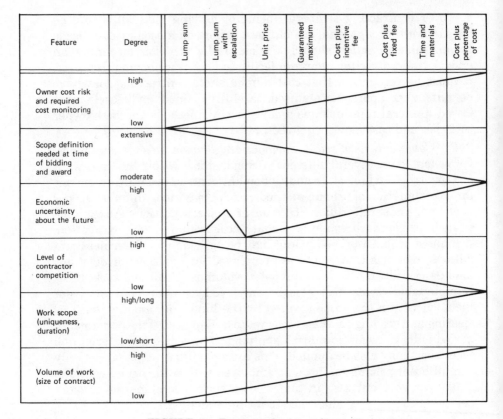

**FIGURE 7.** Features of pricing alternatives.

for performance must be thoroughly specified and understood by all affected parties. And careful monitoring of performance, progress, contractual compliance, and cost data must be ensured. As described throughout this chapter, the degree to which these activities must be performed varies with the pricing method employed.

5. *Owners are not always free to select the pricing method.* Aside from the desires of the owner, there are a number of outside factors or influences that often affect the selection of pricing methods. These include:

*Status of Scope of Work and Performance Information at Time of Bid and Award.* In general, the less defined either the scope of work or performance conditions, such as design drawings and specifications, are when contractual agreements are made, the less chance an owner has of employing fixed-price methods. This relationship is shown in Figure 7.

*Increased Price Inflation or Economic Uncertainty.* The more uncertain contractors (and their subcontractors and suppliers) are about future factors that may influence their performance and costs, the more reluctant they are to agree to fixed-price terms (see Figure 7).

*Scope of Work and Performance Duration.* Longer performance periods (time needed by the contractor to buy, fabricate, deliver, install, or otherwise perform its contractual obligations) and greater scopes of work generally restrict the use of fixed-price methods. This is primarily due to uncertainty or risk of adverse economic trends, such as inflation, or renewal of labor contracts, over long periods of time. Shorter performance periods provide contractors more confidence that factors adversely influencing their profitability will have less effect (see Figure 7).

*Available Sources of Supply.* The greater the number of acceptable vendors or contractors available to the owner, the more freedom the owner has to select pricing methods and contractual terms. Conversely, a small number of acceptable bidders usually reduces the owner's bargaining power, including the selection of desired pricing alternatives. In other words, the environment at the time of contract bidding or negotiation may be either a "buyer's market" or a "seller's market." To the extent that market factors encourage competition among sellers, the owner's ability to dictate pricing methods is enhanced. This relationship is depicted in Figure 7.

*Uniqueness of Scope.* Fixed-price agreements are more applicable to performance that is easily understood, priced, and conducted. In this regard, standard or relatively simple taks, or those that have been accomplished many times by contractors, lend themselves to fixed pricing. Unique, advanced, unproven technology and advanced practices, when included within a contractor's scope of work, reduce the applicability of fixed-price terms. Under unique or unproven conditions, cost-reimbursable methods are generally employed. This relationship is shown in Figure 7.

6. *Subtier pricing need not follow prime contract pricing.* Regardless of the pricing method(s) used in owner–contractor agreements, the contractor's agreements with his suppliers and subcontractors may be different. It is true, however, that subtier agreements may influence

a contractor's preferred pricing method within an owner contract. For example, contractors who can arrange only cost-reimbursable subcontract or supply quotations are generally reluctant to offer owners fixed-price bids. By the same token, owners using cost-reimbursable pricing terms may have some say in future subtier pricing decisions. Although there are cases where subtier and prime contract pricing methods affect each other, the relationships are not direct; nor are they mandatory.

7. *Services are generally more difficult to price than goods.* Because of the inherent variables associated with providing services, they are generally more difficult to price than goods alone. This is particularly true when comparing pricing methods prevalent in the purchasing of material and equipment—concrete, structural steel, cable, electric motors, and so on—and those employed for service or construction contracts—design services, furnish and install electrical instruments, piping installation, and so on. Because the manufacture and delivery of goods typically takes place under controlled conditions (a factory rather than a construction site) and has generally been performed before (mass assembly versus one-of-a-kind installation), the purchase of these goods is more conducive to fixed-price methods. Design or construction services, on the other hand, are often influenced by: (1) greater owner involvement and owner-directed change; (2) environmental factors, such as weather, access, manpower availability, and restrictions in work space and methods; (3) transient, temporary work forces; (4) mobile production facilities, such as cranes, welding machines, and earth-moving equipment; and (5) uniqueness of scope. For these reasons, fixed-price methods are more prevalent when buying material and equipment (purchases) than when buying services (A–E, construction contractors). However, the presence of other previously described factors, such as economic uncertainty, and uniqueness of scope, also reduces the applicability of fixed-price methods within the context of purchase-only agreements.

## CONTRACTOR PERSPECTIVE

Contract pricing methods represent a clear example of differing owner and contractor perspectives, and for very real and understandable reasons, for the basic concept of pricing is that it represents an attempt to share risks between the two. In the case of cost risk, for example, virtually every statement, suggestion, or conclusion drawn in this chapter should be reversed to indicate a contractor's perspective. Where fixed-price contracting generally provides an owner the most protection from cost risk, it simultaneously exposes the contractor to that same risk.

The same relationship is true regarding the required level of owner involvement in monitoring and managing the work. Cost-reimbursable pricing demands close owner scrutiny of contractor efficiency and purchasing practices. Most contractors scrutinize these same practices more closely when they are operating under fixed-price terms—because cost savings under these conditions accrue solely to them.

Because contractors are in the business to make a profit, the concept of cost risk can also be viewed as "profit opportunity." Given a choice, many contractors would probably prefer a well-defined and comfortably priced lump-sum contract over a cost-reimbursable one in which percentage markups for profit and overhead are small. If a contractor feels the risks of overrun are minor with the lump-sum arrangement, it stands to reap more profit than with the cost-reimbursable contract. When the following pricing alternatives are reviewed, it should be noted that although they are described from the owner's perspective, much information may also be derived by considering the contractor's point of view.

## PRICING ALTERNATIVES

Each principal pricing method is described in the following subsections, beginning with firm-fixed-price alternatives and continuing through those classified as cost reimbursable. Absent all other considerations (and there are many), the alternatives are listed in general order of preference from an owner's perspective.

### LUMP SUM

A definite and fixed price is agreed upon prior to award of the contract. This price remains firm for the life of the contract and is not subject to adjustment except for changes in the anticipated scope of work or performance conditions and owner-ordered extras. Under lump-sum agreements, cost risk to the owner is minimal—given adequate bidding and performance controls. Major risks associated with lump-sum contracting include:

1. *The Possibility of Large "Contingency" Amounts Being Included in the Quoted or Contractual Price.* Knowing that there will be no changes in the contract price (except for changes to the work), contractors may artificially inflate the lump-sum amount to cover real or imagined cost risk to them.

2. *Changes to the Work.* Since these are not included in the lump-sum amount, they are priced separately and represent additional costs.

3. *Breach of Contract.* Should the contractor grossly underestimate his costs and therefore agree to an unreasonably low lump-sum price, he may fail to perform the work should his costs approach or exceed the fixed price. This may be unintentional—for example, the contractor loses so much money by performing the work that he goes bankrupt—or intentional—for example, the contractor ceases work in order to concentrate on more profitable jobs. In either case, the owner suffers. Despite that the contract may be legally enforceable, the owner faces schedule delays, quality of work problems, litigation costs, and other detrimental effects.

4. *Reduction of Bidder Competition.* Strict adherence to fixed pricing when other methods are more appropriate could reduce the number of contractors wishing to undertake the work. As a result, the owner would lose the beneficial effects of contractor competition.

There are three major ways in which owners can reduce or mitigate the cost risks associated with lump-sum pricing. They are:

1. Thorough bidder qualification procedures that subject potential contractors to strict financial, quality, and performance standards, which should eliminate unstable contractors, wild bidders, and those without the financial resources to perform
2. Thorough definition of scope prior to bidding and award
3. Control of scope changes and extra work

Considering the above, we see that lump-sum pricing is applicable to most short-duration, well-defined scopes of work. It is most commonly used for the purchase of off-the-shelf equipment or materials and short-duration, smaller construction contracts. It is rarely used when hiring engineering or design services, consulting services, or long-term, large-scope contracting.

## UNIT PRICES

Unit pricing represents a variation of the lump-sum method. Whereas lump-sum pricing involves one fixed price for all, or portions of, the work, unit pricing fixes only the price of a given unit or element of quantity. The total contract price is determined by multiplying unit prices by the quantity of items delivered, erected, or installed. An example of unit pricing could be a contract for concrete placement. The unit price may be $60 per cubic yard, installed. If 1,000 cubic yards were placed, the total contract price would be $60,000.

Owner risk associated with unit pricing includes many of those mentioned for lump-sum arrangements. In addition, unit pricing requires close monitoring and verification of the *quantity* of units actually furnished or installed.

## LUMP SUM WITH ESCALATION

Escalation provisions are commonly applied to both lump-sum and unit pricing. The basic intention of providing for escalation is to eliminate unwarranted price increases to cover contingency in the contractor's bid. Escalation provisions typically call for adjustment based on:

1.  The actual experience of the particular contractor, such as how that contractor's cost changed from an anticipated or base amount
2.  Published indexes of price changes for particular elements, such as labor or certain materials

The use of published indexes is generally more effective when determining additional pricing due to escalation. This is mainly due to the ease with which escalation amounts may be determined. However, when this pricing alternative is used, the following should be considered before an award is made.

1.  *Selection of Appropriate Indexes.*   There are many available, some of which may be influenced by factors not experienced locally, and therefore inappropriate.
2.  *Identification of That Portion of the Base Price to Which Escalation Will Apply.*   Often only certain portions of scope are subject to escalation provisions. These should be carefully defined.
3.  *Application of the Indexes to Progress Payments, Planned Expenditures, and Actual Expenditures.*   This may prove a complex and difficult task.

Escalation provisions may also apply, and often do, to unit prices as well as lump-sum prices. They have become quite common during periods of frequent price fluctuation and for projects that require several years to construct. It is much easier for owners to secure this type of pricing than to secure strict lump-sum or unit prices during such times.

Cost risks to the owner are identical to those experienced under lump-sum or unit pricing, with the following exceptions:

1. Because escalation provisions protect contractors from potential cost increases, contingency amounts included in bid prices are generally reduced.
2. Price increases as a result of general inflation and/or increases in the cost of specific goods or services are borne by the owner rather than by the contractor.
3. Escalation provisions may overcompensate the contractor beyond his actual experience (particularly when based on indexes).
4. Quite often escalation amounts are erroneously calculated, resulting in overpayment or payments that duplicate those made for the base (lump-sum or unit-priced) amount. Escalation provisions should therefore be drawn carefully, with consideration for potential misunderstanding or intentional manipulation.

In order to reduce these risks, owners should insist on clear definition and full enforcement of the items to be escalated, the method of determining escalation amounts, and the processes used for payment of amounts due to escalation. For these reasons, the presence of escalation provisions requires more owner participation during the contract formation phase (determining escalation provisions) and during the performance period (verifying escalation charges) than required for non-escalated fixed-price agreements.

In view of the above, escalation is generally restricted to long-term construction contracts, those involving long lead time equipment or material, and work using cost elements that are subject to higher than average price fluctuation, such as petroleum products.

## FIXED PRICE INCENTIVE

Fixed-price-incentive pricing (also called guaranteed maximum or guaranteed max) allows the fee portion of the price to be adjusted (upward or downward) depending on certain factors that are under the control of the contractor. This adjustment is usually based upon actual costs incurred by the contractor during the work. The following factors are agreed upon and included in the terms of the contract:

1. A *target cost* for the work
2. A *target fee* for the work
3. A *target price* (target cost plus target fee)
4. A *ceiling price* that limits the buyer's responsibility for cost overruns, which is the guaranteed max

5.  A *formula* for establishing the final contractor's fee and therefore
    the ultimate price to the owner, which is called the sharing formula

The relationships among these items for three typical contract applications are depicted in Figures 8, 9, and 10.

Fixed-price-incentive arrangements are used to provide a profit incentive to the contractor to reduce costs of its performance. They do so
by providing a profit-sharing formula under which both the owner and
contractor share in any reduction in costs and, conversely, in any burden
of increased costs. A broad range of incentives are possible. This pricing
alternative is increasingly coming into use in commercial contracts.

Major cost risks to the owner under such arrangements stem from
two sources:

1.  Setting of the guaranteed maximum (or ceiling price). Should this
    be set unduly high, there is little protection for the owner.
2.  Selection of the cost-sharing formula. If the share of cost overruns
    is such that the owner assumes a major portion, there is less incentive
    for the contractor to maintain costs below the target cost.

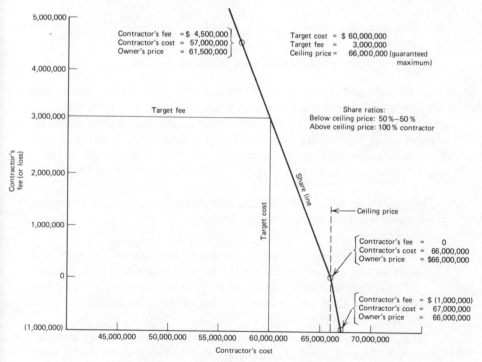

**FIGURE 8.**  Fixed price incentive fee graph No. 1.

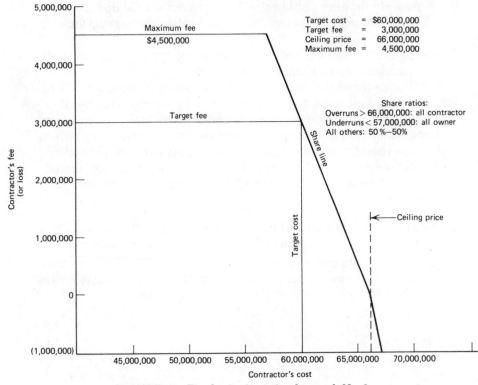

**FIGURE 9.** Fixed price incentive fee graph No. 2.

These risk areas may be reduced through selection of the right target cost, target price, ceiling price, and formula for determining the final price (cost-sharing formula). Many owners, however, experience difficulty in monitoring the ceiling price as changes to the work occur. A major effort is usually required to determine the impact on the guaranteed max amount once scope changes, schedule delays, or contractor claims arise.

The following recommendations are given for owners considering guaranteed maximum arrangements:

1.  Perform an independent estimate of the likely cost of the work and use it to evaluate or verify contractor estimates of the target cost and ceiling price. Make every effort to determine a target cost and ceiling price that are reasonable and contain significant cost incentives for the contractor.
2.  Negotiate a meaningful and realistic target fee and sharing formula. A share formula that is heavily slanted in favor of the owner—for

example, owner receives 95% of underruns, and contractor pays 95% of the overruns—has the effect of turning this contract into a virtual lump-sum arrangement. If the share ratios heavily favor the contractor—for example, contractor receives 95% of underruns, and owner pays 95% of overruns—the contractor has little incentive to reduce costs once the target cost has been reached.

3. Maintain quality standards. Contractors may reduce the quality of their work in order to cut costs and receive higher fees. Constant monitoring of quality should be maintained.

Although this pricing method requires less scope definition prior to award (as compared to a simple lump-sum arrangement), considerable owner involvement is necessary to establish the cost parameters required for an agreement. The setting of the target fee and ceiling price and the formula for sharing cost overruns and underruns are generally the subjects of prolonged negotiation. Once the performance period begins, constant monitoring of changes and the determination of their impact on the target cost and ceiling price are necessary; in many cases, these also involve some form of negotiation.

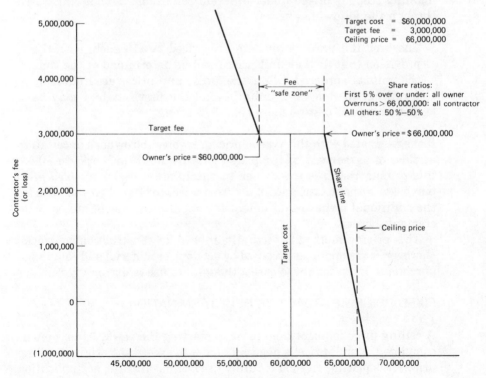

**FIGURE 10.** Fixed price incentive fee graph No. 3.

Guaranteed maximum pricing is applicable for goods and/or services that are difficult to price in a firm-fixed manner. It should be considered, however, only for those contracts where a reasonable estimate of probable costs may be established. Service contracts (such as with an A–E for engineering and design services) may also be based on this type of pricing structure. These contracts often use "target labor hours" rather than a target cost, with the fee adjusted depending on the number of labor hours used. Guaranteed maximum contracts are also common for construction but usually apply to large general contracts rather than to smaller, specialty types of work.

## FIXED PRICE PROSPECTIVE PRICE REDETERMINATION

A firm fixed price is established for a portion of the work, and provisions are made for price redetermination for future work of the same type. For example, the first portion of the contractor's scope may be priced as lump sum or unit price, with these prices agreed upon before work starts. However, once that designated portion of scope is finished, remaining scope is priced under different conditions. This is illustrated through the following example:

> The first 100 units of an item are priced at $15 each, with the provision that further unit prices will be determined at the time the units are ordered. Often, the future unit prices are limited by a ceiling. In this case, units beyond the first hundred may be guaranteed to cost no more than $20 each.

Risks associated with this type of pricing involve the owner's uncertainty at time of agreement of the price of future items or work. In effect, this pricing requires the owner to enter into a series of fixed-price purchases without assurance of the fixed prices to be applied and includes the additional owner requirement to evaluate and negotiate each incremental order.

This pricing method is virtually unused for construction contracts. However, it is sometimes proposed by vendors for material and equipment purchases. Even for these cases, though, its use is extremely rare.

## FIXED PRICE RETROACTIVE REDETERMINATION

A ceiling price is negotiated prior to starting the work. After work is completed, a final price is negotiated, not to exceed the original ceiling price. This pricing method is rarely used. It has little or no application to construction projects and for that reason need not be discussed further.

## COST REIMBURSABLE NO FEE

Under this arrangement the contractor is reimbursed for incurred costs only; no fee is paid. This alternative has little application in the world of commercial construction. When used, it is generally restricted to work performed for nonprofit organizations (universities, charitable foundations, and so on) or for research and development work in which the contractor gains other benefits by participating in the effort. A first-time construction or manufacturing effort where the contractor gains by testing his product or work methods may lend itself to such an agreement.

## COST SHARING

Both contractor and owner share the cost of performing the work. As with the previous alternative, this approach provides no fee for the contractor. In addition, the contractor is responsible for absorbing some of the costs of performing the work. It has little or no application to construction projects. Again, its use is generally restricted to nonprofit work, research projects, or perhaps to the case of construction of facilities that will be shared for the benefit of both owner and contractor when completed.

## COST PLUS INCENTIVE FEE

This pricing method is similar to a fixed-price-incentive-fee (guaranteed-maximum) scheme, except that the contractor is normally reimbursed for all his costs at a minimum. In some cases, however, cost beyond a certain amount may not be fully reimbursed. At that time the contractor may experience a loss (represented by a negative fee). The following factors are determined by the contracting parties:

1. Target cost
2. Target fee
3. Minimum fee $\Big\}$ optional
4. Maximum fee
5. Fee adjustment formula

After performance is completed, actual costs are compared to the target cost, and the contractor's fee is determined through the fee adjustment formula. The general relationship among these factors for a typical contract is shown in Figure 11.

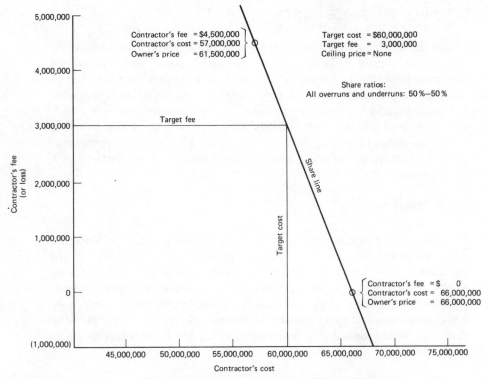

**FIGURE 11.** Cost plus incentive fee graph.

Cost risks to owners are similar to the risks associated with the guaranteed maximum method. Additional risks arise when costs increase beyond the minimum fee level; then, the owner assumes all costs, even though the contractor does not gain additional fees. More detailed owner cost monitoring is required due to the absence of a ceiling price. Extensive effort is needed to set reasonable cost factors prior to award and, as with all cost-reimbursable methods of pricing, postaward monitoring and cost verification are essential.

This type of pricing structure is quite common in agreements with A–E firms and other consultants. As with the guaranteed max scheme, labor hours are often substituted for dollars when determining the target costs (target labor hours). Advantages of this pricing method are:

1.  The contractor has some incentive to keep costs, or labor hours, down.

2. Scope of work can be less specific at time of award than it is in firm-fixed pricing.

3. Agreement may be reached more rapidly than with fixed-price methods.

4. Unnecessary contingencies in bid amounts should be eliminated.

A typical agreement for engineering and design services may use this method in the following manner.

> For the first 600,000 labor hours, the fee will be 80% of labor hour costs; for the next 250,000 labor hours, the fee will be 50% of labor hour costs; for the next 100,000 labor hours, the fee will be 25% of labor hour costs; and for the remaining labor hours, there will be no fee; that is, any time spent beyond 950,000 labor hours will be reimbursed at actual cost—no fee will be earned for those additional hours.

## COST PLUS AWARDED FEE

Under this concept, the contractor is reimbursed for all costs and receives a guaranteed minimum fee plus an "awarded" fee. The awarded fee is intended as an incentive, even though its size is unknown at time of agreement. The awarded fee is subjectively determined after the work has been performed. Again, this subjectivity sometimes includes negotiation.

The owner's cost risk for this pricing method is similar to that found under cost-plus-fixed-fee arrangements (discussed later) because the minimum fee represents a fixed amount. But since the awarded fee is based on a subjective decision by the owner, a contractor's dissatisfaction with that amount may jeopardize future agreements. Oftentimes the awarded fee is viewed as so uncertain as to provide little or no incentive. Establishment of the minimum fee represents the major cost-control effort during the contract formation period. And as for all cost-reimbursable methods, selection of qualified contractors and constant cost monitoring are required.

This type of pricing is used when uncertainty of work scope is too high to establish a ceiling price, or when the work is a first-time attempt and the degree of performance quality is not ensured. Seldom used for furnishing of goods or for construction services, cost-plus-awarded-fee arrangements have been used when contracting for professional services. Even under these conditions, it has limited application.

## COST PLUS FIXED FEE

This alternative allows the contractor to be reimbursed for all costs and to receive a fixed fee for its services. The fixed fee is determined prior to award of the contract and is not changed unless the scope of work is revised. Since it will be reimbursed for all costs, without limit, and its fee is guaranteed, the contractor has no immediate price or cost incentives.

Because of the absence of ceiling prices, awarded or incentive fees, or cost-sharing formulas, there is less effort required on the part of the owner to negotiate or agree with a prospective contractor prior to award. However, some evaluation is required in order to establish the fixed fee. Constant cost monitoring and control on the part of the owner are necessary.

The fixed-fee amount should bear some relation to the degree of difficulty in performing the work, the cost of the work, and the duration of performance. And contractual agreements using this pricing method should specify the required amount of management, supervision, and control provided by the contractor. In addition, the definition of those costs that will be reimbursed and those that are assumed to be covered by the fixed fee should be thorough. Changes to the work require detailed evaluation and negotiation as to their impact on the fixed fee. This pricing method is quite common for construction contracts and for some service-related contracts as well.

## TIME AND MATERIALS

Under a time-and-materials (T and M) contract, the contractor is reimbursed for all materials used at their actual cost. In addition, it is paid for direct labor at a fixed rate (usually hourly). The fixed rates used for labor vary according to labor classification and include a markup for the contractor's fee. As such, direct labor hours represent the only vehicle for contractor profit and recovery of nonmaterial costs. The hourly rates should include wages, overhead, general and administrative expenses, and profit.

The time-and-materials pricing method provides no incentive for controlling material costs or labor costs. In addition, it may be argued that it gives the contractor an incentive to increase direct labor hours used because as these increase, its embedded fee increases proportionately.

Contracts or change orders based on the T and M pricing method are simple to obtain and can be awarded rapidly. Therefore, they are commonly used for small scopes of work, such as that needed for repair or emergencies. Greater effort is required on the part of the owner in monitoring both material costs and labor hours usage. This method is quite common for construction work and for professional services. In general, it should be avoided for all work except that involving minor-cost, short-duration efforts that cannot be planned in advance.

## COST PLUS PERCENTAGE OF COSTS

Under this arrangement, contractors are reimbursed for all costs and given a fee that is directly proportional to some or all of the costs involved in the work. The higher the direct costs are, the higher the contractor's fee is. Most owners agree that this arrangement is the least favorable of all available pricing methods. Not only do contractors have no incentive to control costs, but they also are given a positive incentive to increase costs. Because of the lack of cost control offered by this pricing arrangement, the owner's efforts during selection of the contractor and subsequent monitoring of costs and its cost-control efforts should be substantial. In many cases the fee percentage is renegotiated during the performance period, particularly if incurred costs have exceeded original estimates. This also requires considerable time and effort. Perhaps the best control recommendation regarding the cost-plus-percentage-of-costs approach is to avoid this pricing method. If it is unavoidable, efforts should be made to set the lowest possible fee percentage and to monitor both costs and performance of the contractor very carefully. Cost-plus-percentage-of-costs arrangements are not recommended for any application. If they are used, they should be restricted to short-duration, low-cost, or emergency work performed by reliable contractors.

## PROFIT INCENTIVES

Cost incentives described for several of the preceding cost-plus pricing methods represent only one form of contractor incentive. In addition to target prices, ceiling prices, share ratios, and the like, other inducements—both positive and negative—are used to promote efficiency, quality, and timeliness of contractor performance. The available incentives are grouped into the categories of cost, technical, and schedule, according to the element of performance they are intended to promote. Taken together, they are all thought of as contractor-profit incentives.

## TECHNICAL INCENTIVES

Technical incentives punish or reward contractors for the quality or performance of work. They are commonly used when certain performance characteristics of a finished product are important or valuable to the owner. The concept of rewarding or penalizing a contractor according to the performance of a finished product can be applied to most contracts where design and development work is involved in addition to construction. A general principle is that it does little good to create incentives for performance that is beyond the construction contractor's control. Performance of a cooling tower, for example, is greatly dependent upon the design and manufacture of its components. To penalize or reward a contractor responsible for only the installation of the tower would make little sense.

## SCHEDULE INCENTIVES

The term *bonus and penalty clause* usually refers to schedule incentives—sometimes called delivery incentives. These seek to reward contractors for early completion of work or penalize them for late completion. In some cases, only penalties are involved. These penalties are often referred to as liquidated damages and are typically stated in terms of so much money per day beyond a specified date. They have merit for those projects where completion by a certain date is essential, or where significant costs are saved, in the case of early performance, or incurred, in the case of late completion, on the part of the owner.

In combining cost, schedule, and/or technical incentives into a contractual agreement, owners must ensure that the different incentives do not counteract or conflict with one another. For example, a contractor absorbing 25% of cost overruns beyond a certain target price in addition to gaining $250,000 per day for early completion would be inclined to spend tremendous amounts of money—on overtime, extra equipment, prefabricated materials, and so on—in order to secure the lucrative schedule incentive. In doing so, it would naturally forgo the 25% increase in required cost. This trade-off may be contrary to the interests of the owner. All types of incentives—cost, schedule, and technical—should be balanced when used together on a single contract. The same advice applies equally to situations where more than one contractor is working on a project. Should an earlier contractor be working under extremely high schedule incentives, it may rush its work at the expense of the interfacing contractor who follows. Many times owners adopt incentive arrangements without considering the accompanying direct and indirect

effects on the contractor in question as well as others involved with
the project.

## PRICING CHANGE ORDERS

Regardless of the pricing techniques used for the original or "base"
contract, construction contract documents invariably provide for the
pricing of changes or additional work according to one of several prearranged methods. Typically, the owner allows itself the option of requiring
the contractor to quote on change order work using one of three methods:

1.  An agreed lump sum
2.  Unit prices already contained in the contract or subsequently agreed
    to between the owner and contractor
3.  A cost-plus-percentage fee, with the definition of allowable costs
    and the acceptable fee percentages already stipulated in the contract

Some owners specify the percentage fee for the third method, above,
when designing the bidding or contract documents. Others, however,
ask the bidders to insert the fee they will charge for change-order work,
and, depending on the amount of extra work anticipated, this could
become an important factor in evaluating bids and awarding the contract.
Rather than using only one percentage markup amount, most contracts
call for different markups depending upon the element of cost involved.
For example, a contract might call for the following markups to cover
the contractor's overhead and profit should extra work be ordered on
a cost-plus basis.

15% for direct labor
10% for material
8% for subcontracted work

Again, the owner usually retains the right to choose the pricing method,
one of the three described above, by which the contractor must quote
for additional or extra work. If the contract is unit priced and additional
materials or labor are required, the existing unit prices may cover
these additions up to a certain amount (say, 25% above the anticipated
or estimated quantities). Some contracts require that any amount over
or under such a range will be quoted on different unit prices.

Even with lump-sum contracts, it is wise to request bidders to quote
unit prices for additions or deletions to the work should changes occur.

By requesting these prices with the original contract bids, the owner receives competitive prices before it awards the contract. In their absence, it could find itself in a poor negotiating position later—when the contractor is on site and has the owner "captive." Items that lend themselves to "add-and-delete" unit pricing include tons of steel, cubic yards of earthwork and concrete, square feet of road surface, and feet of electrical cable and piping. It is not uncommon to see contractors bidding a different unit price for adding versus deleting the same items. For example, a bidder may quote an "add " price of $1.10 per pound for steel reinforcing bar and a delete price (or credit) of only $0.90 per pound for the same bar. This is an accepted practice and is based on the fact that certain fixed costs and overheads for the contractor do not decrease in direct proportion to the amount of reinforcing steel installed.

## SUMMARY

The selection of contract type and associated pricing strategies should be a logical process that considers the nature of the work, time of performance, contractor marketplace, completeness of design information at time of bid, and the resources of the owner to monitor and administer the contract. It goes without saying that more contract monitoring is required for a cost-plus agreement than for a lump-sum one. And regardless of the owner's wishes, it may have to settle for a different pricing arrangement if no contractors will accept the one it has chosen. Also, all contracts should have specific provisions for pricing extra work. It is preferable to include these as items to be bid at the time of original contract formation so as to take advantage of the owner's leveraged position.

One final note: it is surprising, considering all of the above, to see the number of owners who choose to ignore these considerations and select one broad pricing structure to cover all contracts on a large project—for example, "We use only lump-sum contracts." Oftentimes this is an arbitrary, illogical decision based on personal bias and inaccurate assumptions. Each contract is different, and a pricing strategy tailored to the risks associated with each particular contract is simply good business practice.

**CHAPTER 4**

# CONTRACT PACKAGING AND SCHEDULING

The first task the person responsible for contract formation (we shall call him or her the contract administrator for expediency) must undertake at the beginning of a proposed construction project is to determine the orderly subdivision of the work into distinct contracts, or preliminary contract "packages." Basic decisions must be made concerning: (1) the number of contracts to be used; (2) their respective scopes of work; and (3) schedule requirements for their bidding and award.

Contract packaging efforts, of course, depend directly on the organizational approach chosen for the project, such as design–build, force account, and multiple primes. Where a single design–build contract is to be used, this packaging and scheduling process is reduced to a simple and straightforward exercise—there is only one contract involved. With the opposite extreme, in a multiple-prime environment, this process can be extremely complex and difficult. And, naturally, there are gradations in between these two examples. For purposes of presenting an exhaustive analysis of this topic, we will use the multiple-prime approach. It will present us with the most challenging and thorough illustration.

## PRELIMINARY SCOPE DETERMINATION

This is the project management team's first attempt to "scope out" or define the work that each contract will represent. It should be a joint effort of personnel representing construction, engineering, and contract administration, as a minimum. The results of this preliminary discussion should be: (1) preliminary scope documents for each anticipated contract;

and (2) a milestone schedule indicating at least the following dates for each contract:

1.  Contractor commences work
2.  Contract award
3.  Bids due
4.  RFP released for bids
5.  Delivery of design products (drawings and specifications)

## CONTRACT SCHEDULING

The preceding milestone dates are listed in reverse order of occurrence for a significant reason. They illustrate that the integrated project schedule should be construction driven; that is, the contract formation and engineering functions should be scheduled and performed in order to support the needs of construction. This represents a simple "backward-looking" schedule philosophy, and the intervals between these milestones are the durations allowed for contract formation activities. This can best be shown by referring to the abbreviated critical path method (CPM) network, depicted in Figure 12. In order to obtain challenging but achievable schedules, the contract administrator must allow sufficient time for the major tasks required during the formation period.

When preliminary contract packaging and scheduling is complete, engineering will be able to plan its work in order to support the formation and construction processes because it will know:

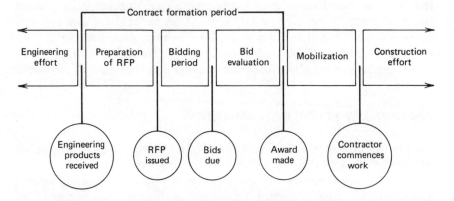

FIGURE 12.   Contract formation milestones and activities.

1.  The priority, sequencing, and grouping of its technical products
2.  The delivery dates required for those products

Contract administration will know the number and scope of the required contracts, the due dates and durations for the formation tasks associated with each, and the due dates for contract award so that construction may proceed as planned. Construction personnel will know the number and order of mobilization of the project contractors, as well as what each one is expected to perform.

## ITERATIVE PROCESS

This scoping and scheduling process should not be thought of as a one-time effort. On the contrary, flexibility and changes will require that it be iterative and that the results constantly be updated for each contract, with the project schedule reflecting the latest dates and durations. An example of a preliminary scoping document for a particular contract—site preparation for an electric power plant—is given as Exhibit 1. It is extremely helpful for those planning the contracting process to develop such a document for each envisioned contract. When used, each contract should be defined as to scope of work, suggested pricing method, required engineering products, and closely related work not included.

This documentation is an extremely valuable communication tool. It tells the engineering group (or consultant) what support information is required for each package, allows the contract administration and construction groups to verify that all work is covered by one and only one contract—no gaps or overlaps—and allows the owner's construction organization to load its required field monitoring and support services, such as staff, field offices, craft parking lots, access, and sanitary facilities, over the project construction phase. In effect, the collection of planning documents for each contract becomes the owner's "contracting plan."

When all contracts have been preliminarily scheduled, the contract administrator can determine the level of effort required at any given time in order to perform contract formation functions. This allows the contract administrator (or an immediate supervisor) to assign the appropriate manpower and to attempt to level this manpower over a period of time—to reduce peaks and valleys—if possible. A typical CPM network segment showing the durations and relationships of the major contract formation activities they will perform is shown in Figure 13.

**EXHIBIT 1. Contract Packaging Scope Document**

CONTRACT PACKAGING SCOPE DOCUMENT

**Contract Identification**   C–11   Site preparation

**General Description of Work**

Preparation of construction site, including clearing and grubbing, grading, cutting and filling, installation of storm drainage and erosion prevention systems, construction of sedimentation pond, access roads and parking lot subgrades, preparation of material laydown area, and performance of initial slope seeding.

**A.**  Scope of Work

1.  Clearing and grubbing site vegetation
2.  Excavation, grading, and soil compaction
3.  Preparation of fill storage and disposal areas
4.  Transfer of soil, rock, and other excavated material to storage and disposal areas
5.  Reclamation of excavated material for use as fill
6.  Transfer and disposal of waste material to land fill area
7.  Hauling and placing of fill material
8.  Local dewatering of contractor trenches and other excavations
9.  Initial seeding of slopes where required
10.  Installation of storm drainage pipe, prefabricated culverts, and catch basins (to be furnished by owner)

**B.**  Technical Specifications Required

No. 122 Earthwork
No. 145 Site drainage system
No. 232 Landscaping

**C.** Drawings Required

    1. Layout:
       Site plot plans
    2. Civil:
       Mapping and property
       Subsurface exploration
       Site preparation
       Sanitary waste system
       Dams and reservoirs
       Site improvements
       Yard structures and buildings

**D.** Closely Related Work Not Included

    1. Excavation of foundations for temporary or permanent buildings
    2. Concrete construction, including material supply
    3. Furnishing of drainage pipe, culverts, or catch basins
    4. General site dewatering (not including contractor excavations)
    5. Installation of subgrade piping (except for storm drainage system)
    6. Construction of roads, parking, or laydown areas above subgrade

**E.** Recommended Pricing Structure

    1. Lump sum (firm fixed price without escalation)
    2. Unit prices for addition and deletion of quantities of:
       rock ⎫
           ⎬ excavation, fill, disposal
       soil ⎭

**F.** Contract Formation Dates

    1. Begin construction June 1, 19XX
    2. Award contract May 1, 19XX
    3. Receive bids March 15, 19XX
    4. Release RFP February 1, 19XX
    5. Receive drawings and specifications January 10, 19XX

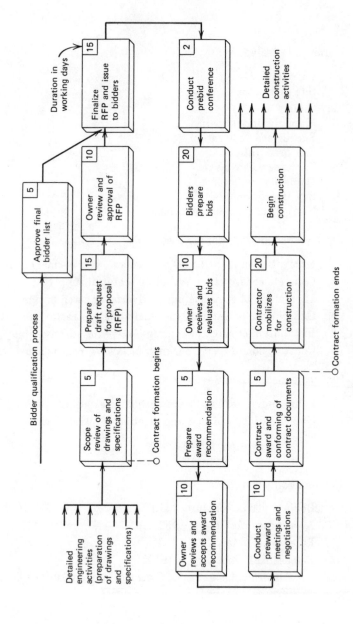

**FIGURE 13.** Sample contract formation CPM network.

## INTEGRATED SCHEDULE

It should be noted that the contract formation activities represent the locus of intersection of the preceding engineering–design effort and the subsequent construction effort. This planning forces these two, sometimes disparate, schedules to join in an overall integrated project schedule. The contract packages themselves form a series of time-phased bridges between these two schedules. Sufficient planning of each formation step and use of reasonable durations for each activity help prevent these bridges from becoming bottlenecks.

## CONTRACT IDENTIFICATION

It is customary to number each contract for ease of reference and tracking purposes. Most organizations will simply number them chronologically beginning with the first to be awarded and continuing to the final contract (C–1, C–2, C–3, and so on). Often prefixes will be attached to the numbers to indicate engineering discipline or construction trades associated with each contract. For example, two electrical contracts are called E–1 and E–2; mechanical contracts, M–1, M–2, M–3, and so on. Long, convoluted multicharacter identifications are strongly discouraged, for each will be used thousands of times throughout the contracting process.

## REPRESENTATIVE CASES

1.   A project for constructing an 80-story office tower and adjacent parking garage was in process. A multiple-prime contracting approach was being used, and the issuance of the RFP for furnishing and erecting structural steel was being delayed because the steel design drawings were incomplete. After several weeks, the lead structural engineer, tired of being made a scapegoat, exclaimed that he was being held up by the indecision on the part of the owner's building manager regarding the choice of structural steel versus precast concrete for the elevated parking decks. In actuality, the parking garage was never intended as part of the scope of work for the structural steel contractor—even if it were to be made of steel! The garage was singled out as a separate turnkey contract much earlier, but this decision was unknown to the lead structural engineer.

2.   The chief of contract administration reviewed the integrated project schedule and decided it was ambitious but achievable. Durations given for the formation of each contract were reasonable. He failed, however, to consider the combined effect of all contracts on the manpower in his department at any given time. Murphy's Law prevailed, and he later found his three contract administrators scheduled to conduct six prebid conferences, issue four RFPs, and evaluate 12 contract bids during the first week of April! During the remaining three weeks of that month, they had no assigned work, so they spent their time grumbling about their leader and updating their résumés.

3.   Because he knew that a specifications writer was to be hired in several months, the lead mechanical engineer decided to concentrate his staff's efforts on mechanical design and drawing production during the intervening period and to defer the writing of contract specifications until the new man came aboard. The drawings for all five mechanical contracts were completed ahead of schedule, but two FRPs were severely delayed due to lack of technical specifications.

4.   The electrical drafting squad leader made a meticulous takeoff from the master integrated project schedule at the beginning of the project. He created a separate, detailed drawing production schedule that, if followed, would ensure that his drawings would be ready for each contract RFP when needed. During the next several months, work on the chemical plant's generating unit was delayed due to permit problems, the cooling tower was accelerated by 10 weeks to escape a pending environmental deadline, and purchasing lead time reports showed transformers requiring six months for delivery instead of the original three. All this was reflected on the updated project schedule, but since the drafting squad leader had been operating from his personal "pocket schedule," his production priority was totally out of synchronization with project needs.

# PART 2

# CONTRACT FORMATION

The previous chapters have dealt with the assignment of responsibilities and selection of pricing and scheduling controls for an entire project; this part will now address the sequence of steps used to secure fair and equitable contracts that: (1) are well defined and priced; (2) protect the interests of the contracting party; (3) contain enforceable compliance controls; and (4) can be properly administered.

Each of the following chapters describes steps taken in a sequential manner, for each contract, to achieve these objectives. These steps consist of:

*Developing Contract Documents.* The first step is design and selection of project- or company-wide documents that will form the basis of control for each contract-specific application.

*Bidder Qualification and Selection.* This formal, structured process is aimed at reducing bid evaluation effort, soliciting competitive bids from the best possible offerings, and preventing performance problems once construction has started by allowing only responsive and qualified contractors to participate in the bidding process.

*Issuing Requests for Proposals.* The general reference or standard documents are tailored into specific bid packages. Steps taken to prepare each RFP will be reviewed and operational considerations offered to help achieve well-priced, clearly scoped, and tightly bid contracts.

*Managing the Bid Cycle.* The selection of the best bid offering begins long before bids are received. This chapter will describe ways to channel bids into acceptable offerings by control of bid information prior to receipt through precise use of addenda, prebid conferences, bid extensions, and contractor site visits.

*Bid Receipt and Evaluation.* Methods must be followed to control the evaluation process, maintain bid security, and ensure that the cost, technical content, and commercial considerations of each bid are thoroughly analyzed so that the best possible offering is accepted.

*Contract Award.* Once a successful bid has been identified, positive steps can be taken to conform the documents into enforceable and easily administerred criteria. The use and abuse of letters of intent, as well as the incorporation of changes and understandings made during preaward meetings and negotiations will be described.

These steps will take us from the point where the need for contracting for outside services is recognized to the achievement of a signed contract, with a qualified contractor, at acceptable levels of commercial cost and risk. Each of these steps will be performed at various times for each proposed contract.

Acts or omissions during the formation phase greatly affect the success of contract administration and contractor performance to follow. Formation controls are, therefore, important not only because they benefit the selection and award process but also because they set the stage, cast the players, and provide the script by which the actual contract work will be performed.

# CHAPTER 5

# DEVELOPING CONTRACT DOCUMENTS

Two distinct sets of documents are used by owners to communicate their needs and to structure the activities of both parties during the contract formation and administration phases: bidding documents and contract documents. For most complex projects, these are often voluminous in size, encompassing in scope, and very detailed in content. Owners embarking on the formation of construction contracts have three general sources of contract and contract-related—that is, bidding—documents:

1. *What Worked Last Time.* The owner chooses the most appropriate and best documents from its files of previously used documents.
2. *Standard Documents Available in the Industry.* These include bidding and contract documents published by professional associations or certain industry groups involved with construction. Examples are those provided by the American Institute of Architects (AIA), the National Society of Professional Engineers (NSPE), and the Associated General Contractors of America (AGC), to name a few. Standard contract documents are also published by various associations for specific application to international projects.
3. *Reference Documents Maintained by the Owner or Its Representatives.* These include documents prepared and maintained for a range of applications. They usually contain standard clauses and terms and provide for the modification or insertion of project, owner, contract, or jurisdictionally specific terms and conditions. They are often maintained on electronic memory for rapid access, editing, and revision prior to a specific use.

Of these three choices, most owners choose from the final two. They realize that unless "what was used last time" is recent, was effective, and does not differ substantially from present needs, much revision is normally required to modify specific contract documents to a new application.

## INDUSTRY STANDARD VERSUS OWNER REFERENCE DOCUMENTS

Although owners of major projects have used industry standard documents with varying degrees of success, these are much more commonly found in conjunction with smaller commercial or architectural construction. In order to determine whether industry standard or owner reference documents should be used as the starting point for developing specific documents tailored to needs of each proposed contract, owners should consider the benefits and disadvantages of each. This decision is generally reduced to the advantages of standard documents centering upon their acceptability and inexpensive preparation when compared with the more expensive yet more specific reference variety. To assist owners in making this decision, the following advantages and disadvantages of each are suggested:

## INDUSTRY STANDARD DOCUMENTS

These documents have several advantages:

1. They have generally been carefully designed and reflect the collective experience of many practitioners in many project applications. They offer basic controls over common contractual risks. They are updated periodically to incorporate enhancements gained through their use and to reflect changing industry practices and legal interpretations.
2. They can be obtained and used easily. The effort of creating contract documents is greatly reduced, and the time required to achieve owner-specific versions is sometimes minimal.
3. Some proponents of industry standard documents claim that they have more weight in court than those created by the owner (or contractor) without the endorsement of a sponsoring agency or association.
4. They are probably more familiar to contractors. As opposed to specifically designed documents, standard documents have been seen

and used by most major contractors, and contractors are often more comfortable with standard documents during the bidding stage.

5. They are inexpensive to obtain and to use. Most are available without charge.

These standard documents do involve some disadvantages, however:

1. They generally favor the drafter (sponsoring organization). An examination of documents prepared by engineering or architectural associations compared with those prepared by contractor groups points this out very directly. Protection is often afforded the party represented by the sponsoring association at the expense of the others.

2. They reflect assumptions that may not be merited for specific applications. Most industry documents are written to have broad application, in many jurisdictions and under widely varying circumstances. Because of this, they do not contain specific language reflecting owner needs as they may vary from one application to the other. It is difficult to isolate and revise these assumptions to reflect specific owner management and control objectives without an exhaustive analysis. Often the assumptions are implied or embedded in nonrelated clauses.

3. They reflect "typical" or assumed roles and responsibilities. The roles and responsibilities assumed for major contracting parties, such as owner, architect, engineer, contractor, subcontractor, and construction manager, may be very different for a specific project from those assumed by the drafters of standard documents. It is not unfair to say, in addition, that the authority is slanted toward the group represented by the drafting organization, and questions of liability or responsibility are slanted away from that group as well.

## OWNER REFERENCE DOCUMENTS

Owner reference documents have advantages over industry documents:

1. They can contain specifics dealing with owner, project, and jurisdictional settings. They may be designed to be more precise and specific, and therefore more easily applied to the particular contract in question.

2.  They can implement specific control objectives that may not have been envisioned by the drafters of generic documents.
3.  They reflect known rather than assumed roles and responsibilities.
4.  They can contain owner protection not afforded in different jurisdictions. For example, should a hold harmless clause favoring the project owner be illegal in a majority of the states, industry documents will most likely omit this protection. If the owner and project in question are not bound by such legal restrictions, reference documents may incorporate the protection offered by such a clause.
5.  They can match other owner-prepared documents more closely than can generic versions. These include any other bidding and contract-related forms, instruments, and reports (change orders, technical specifications, backcharges, and so on).

However, they have disadvantages as well:

1.  They are more expensive to obtain. Much time and effort is required to create and maintain owner reference documents.
2.  They may contain clauses whose enforceability is untested.
3.  They may be unfamiliar to contractors. Bidders and contractors may be uncomfortable with or suspicious of the terms and conditions proposed by the owner.
4.  They may be poorly written or incomplete. The quality of reference documents is a function of the knowledge and ability of those who prepare them. When owners do not possess the ability to create clear, concise, and enforceable documents they should engage the services of those who do.

The decision to choose one alternative document source over the other must be made by the individual owner with full awareness of the benefits, costs, and risks pertaining to each, and in full view of the specific circumstances surrounding project contracting applications. For most owners of large industrial projects, reference documents are created during the initial planning stages of the work and are revised to reflect the specifics of each anticipated contract. This chapter will assume this to be the case when discussing the documents themselves and the operational considerations involving their use.

Standard documents may be used, but their effectiveness depends upon the owner's understanding of their contents and a careful tailoring of their general terms to match specific needs and control objectives. In any case, both types of documents should strive for consistency and uniformity in their format and content and represent good starting points for contract-specific modification—which is always required, no matter the source of the original documents to be used.

## CONTRACT AND CONTRACT-RELATED DOCUMENTS

Bid packages generally contain most or all of the following types of documents (examples of each are given as exhibits in Chapter 7):

1. *Invitation to Bid.* This is usually in letter form and sent to bidders inviting their proposals and explaining the rules of the bidding process.
2. *Proposal.* This is completed by the bidder with such information as prices, schedules, and submittals. It references the remaining four sections and the drawings as the basis for the contract.
3. *Agreement.* This contains the agreement that will be executed by the contracting parties once a bid is accepted.
4. *General Conditions.* General requirements that apply to all construction contracts for the particular project are given here. The basic relationships between the contracting parties, general project rules, and commercial terms are contained in this section.
5. *Special (Supplementary) Conditions.* Particular requirements of work to be covered under the contract are given here. These include scope of work, material and services furnished by the owner or others, special site conditions affecting access, laydown areas, storage and utilities, and limits of responsibility of the contractor.
6. *Technical Specifications.* Prepared by the responsible A–E organization, the technical specifications give detailed technical requirements of the work to be accomplished. Codes and standards, performance and acceptance criteria, and materials and methods are prescribed by this section. The technical specifications are a group of documents (or divisions), each pertaining to a particular installation or service involved in the work to be performed by the contractor.
7. *Drawings.* Prepared by the responsible A–E organization for the owner, these graphically depict location, size, shape, and details of construction or composition of the work.

The preceding items comprise the bid package or request for proposal (RFP) package delivered to bidders. Once a successful bidder is selected for award, items 1 and 2 are withdrawn, and a conformed contract containing items 3–7 is prepared and executed. These items are called the contract documents (see Figure 14).

## CONTRACT DOCUMENT DESIGN PHILOSOPHY

For smaller commercial and architectual construction projects, many owners and design professionals prefer to adopt standard industry-endorsed bidding and contract documents, such as AIA documents, and modify these to the particular needs of the owner, contractor, and project conditions. In a parallel effort, the Construction Specifications Institute (CSI) has developed detailed technical specifications formats and guidelines. When properly and appropriately used, these documents and guidelines greatly reduce the document creation process, contribute to uniformity among contracts and projects, and result in high-quality documents.

There is much debate, however, among those who deal with large engineered projects—power plants, refineries, transportation systems, heavy industrial facilities, and so on—as to their application in these environments. A general consensus, if one exists, is that without considerable modification the generic, industry standard formats are not especially suited to these types of projects. Those responsible for preparing contract documents (including specifications) in situations where available standard documents are deemed unsuitable are required to design and create their own. Often this is done on a project-by-project basis.

**FIGURE 14.** Typical bidding and contract documents.

However, repeat builders and engineering or project management consultants have recently begun to produce "reference" contract documents that can be modified quite easily to suit the requirements of a particular owner, legal jurisdiction, project type, and organizational–contractual approach.

Regardless of the origin of the bidding and contract documents, there are certain guidelines, or caveats, that should be followed to streamline the processes they affect: contract formation and administration.

## CONTROL OF THE DRAFTING PROCESS

One of the most foolish actions an owner can take is to request proposals from bidders without specifying the format and content it expects to receive. In effect, it is saying, "Bidders, send us whatever you choose." It is easy to imagine the menagerie of different proposals that will surely result. The bid evaluation process is difficult at best, and chaotic as a rule. Comparing widely dissimilar proposals would resemble comparisons among apples, oranges, bananas, and bricks.

Most astute owners (or their representatives), therefore, channel the bidders into conforming to particular proposal formats. This is done by using a properly designed bid package (RFP) and informing prospective bidders that deviations from them may be cause for rejection of the offending bids. A well-designed RFP is simple and direct and completely self-explanatory and reflects a reasonable attempt to avoid misconstruction of its requirements.

RFPs designed to this end contain two major sections: (1) the "bidding only" documents; and (2) specimens of the contract documents that will be used once a successful bidder is selected for award. Bidders should be furnished a detailed proposal form, soliciting specific information, and required to use this form to submit the details of their bids. Use of "bidder's paper"—a bidder's own proposal format, standard conditions, and so on — should be prohibited. Aside from complicating and extending the bid evaluation and contract conformance process, this offense would require the owner's staff to review completely each detail of each bidder's documents for each contract (a lot of fine print to squint at). It cannot be overemphasized that the owner must maintain authorship and control over the bidding and contract documents. From the postaward perspective, contract administration is difficult even with several contracts containing identical commercial and legal terms. Were each of these to be significantly different, the contract administrator would soon become convinced that he or she is working on a project to build the Tower of Babel.

Bidders have a knack for inserting their own clauses, deleting existing ones, and qualifying their bids through other means. Again, to have to check for interlineations, erasures, typeovers, and marginal notes is an unbearable and risky task. Each of these should be prohibited and bidders required to state exceptions to the provided documents on a separate sheet of paper, bearing their letterhead, with suitable references to the specific items as addressed in the owner-furnished documents.

Some owners will award contracts with a letter stating "on the basis of your proposal dated" or, rather than transferring bidder quotations from the proposal form to the agreement form before the contract is signed, will instead incorporate the bidder's proposal into the contract by reference. This is dangerous and should be avoided. It takes little time and effort to transfer the quoted prices and other variable information— names, dates, and so on—to the owner-designed agreement form. The bidder's proposal can be filed away. To facilitate this process, the proposal and agreement forms should be parallel in design.

## OPERATIONAL CONSIDERATIONS

In addition to the general guidelines given above, there are several specific suggestions that will aid the formation process. For example, where certain documents are required for all contracts of a project and are not subject to modification, consider preprinting several hundred copies to save reproduction costs and time. These might include the following:

General conditions
Insurance requirements
Project agreement
Job rules
Safety program requirements
Special instructions (for example, drawing submittal requirements, or quality assurance program).

## DOCUMENT RELATIONSHIPS

Document hierarchy (which documents should dominate if conflicting requirements are discovered, and so on) should be fully described in the documents themselves. A general rule is that written instructions, requirements, and so on, should occur only once and in the proper documents. The drawings are issued with the bid packages (RFPs) to give the bidder (and later as "contract drawings" to give the contractor) such necessary information as can be transmitted more effectively in graphic form than in written language. For this reason, instructions that are able to be written should be placed in the technical specifications rather than in the drawings. Anyone who has been through this process knows how much more difficult, time-consuming, and expensive it is to revise and reissue drawings as opposed to modifying typewritten contract documents when changes occur. This is a pragmatic reason to follow this suggestion. To that end, the people preparing the drawings and those preparing the written documents need to communicate so as to avoid redundancies and conflicts between the two.

This same principle applies to the technical specifications and their relation to the other written documents (general conditions, special conditions, and so on). A common practice is for engineering departments to author the specifications and drawings while contract administration prepares the remainder of the documents (commerical "boiler plate"). Each should be aware of the topics it is to address. Often, the technical specifications contain commercial and legal terms or instructions embedded in the technical requirements. This creates three apparent problems:

1. There is a strong possibility that they are in conflict or redundant with the commercial sections.
2. If technical specifications remain only technical, they can be readily used for other projects or other contracts without reissue or modification. If they contain project-related terms and conditions, this efficiency is forgone.
3. Generally speaking, personnel trained to create technical specifications are not qualified to decide on or address commercial and legal questions.

The same is true for the drawings. Since the same drawing may be used for several separate RFPs and contracts (with each contractor performing only certain portions of the work), the scoping of the work among different contractors should be done through the special conditions section. Notes on drawings such as "by Owner," "by Others," or "by C–15 contractor" are discouraged. These scopes will change as the project progresses, and again, it is much easier to reallocate scope through written documents than to redraw, check, approve, reissue, reproduce, and distribute hundreds of drawings with scores of copies each to several affected contractors and internally among the owner's and designer's organizations.

## USE OF PURCHASE ORDERS

Many owners issue purchase orders to cover virtually all items they buy. Accordingly, they issue a purchase order to procure contractors' services, or contracts. This is acceptable, providing the purchase order does not contain standard or preprinted terms and conditions (most of them do). The problem lies with these terms and conditions. They are intended to cover commercial *purchases*—that is, office supplies, raw materials, fuel supplies, and so on—and do not suit the special requirements of construction contracting.

A basic tenet is that purchases involving job-site labor (with or without the furnishing of material or equipment) should be accomplished through a construction contract and not a purchase order. If owners insist on issuing purchase orders (POs), the PO should merely reference the applicable contract—one that has been prepared and formed in accordance with the recommendations contained in this book—and should specifically state that all printed conditions on its face (or reverse) are null and void.

The use of POs is an acceptable practice, but when used in conjunction with construction contracts, they should be reduced in significance to that of a tracking and transmittal vehicle, not as a replacement or supplement to the contract documents. For the same reasons, change orders to contracts should stand alone, and PO amendments or modifications should not replace or supplement these either.

## DOCUMENT MODIFICATIONS

Changes to contract or bidding documents should take two forms: (1) changes to the bidding documents (RFPs) during the bid cycle (called addenda); and (2) changes to the contract documents once the award is made (called change orders). An intermediate state exists when an accepted proposal is being incorporated into the final contract documents prior to their signing (or execution). No formal instruments are required for this process. Addenda are issued to bidders during the bid cycle (while RFPs are "on the street"). They should be numbered, dated, and tightly controlled. This is also true for change orders. Both will be discussed in detail in later sections.

The point of introducing these two modification instruments here is to distinguish them from each other and to emphasize that in order to maintain control over both RFPs and contracts, strict adherence to these principles should be enforced:

1. RFPs should not be amended unless by addenda.
2. Contracts should not be amended unless by change order.

Some owners accept the use of a plethora of change instruments. These include such things as contract modifications, amendments, extra work orders, field orders, contract guidance and intent statements, and bulletins. Problems in controlling and determining the status of RFPs and contracts and processing contract transactions—to name a few—increase geometrically with the introduction of each additional instrument of change. In order to maintain a disciplined and controlled approach to contract formation and administration, addenda and change orders, when properly used, are all that are needed to implement change.


## ANCILLARY CONTRACTS

There are occasions when the format or content of documents designed for major construction contracts are not well suited. Three major examples are:

1. *Brief-Form Contracts.* These are designed for rapid deployment from a field site location. They cover short-duration, low-cost activities. They should be designed and issued under the same guidelines used for the formation of major construction contracts. But due to the nature of the work (snow removal, temporary access roads, and so on) and for expediency, they are streamlined, condensed versions. Brief-form contracting is discussed in Chapter 16.
2. *Equipment Rental Contracts.* When fully operated and maintained construction equipment is rented, special terms and conditions apply. Bidder selection, proposal submission, and contract preparation are essentially the same as for brief-form contracts.
3. *Technical or Consulting Services Contracts.* Specific terms are applicable for such services as architectural or engineering services, inspection and testing services, and surveying. Topics important to construction contracts are not required, whereas specific terms

relating to patents, professional licenses, and liability, among others, are appropriate.

If a significant amount of any of the above instances is envisioned, it may be advisable to prepare standard or "reference" documents for them as well.


## BONDS

There are two major categories of bonds in use today for commercial transactions and relationships. These are (1) fidelity bonds, which offer protection or indemnification against dishonesty on the part of the principal (bonded person or organization); and (2) surety bonds, which protect against loss due to the principal's failure to perform under a contract or to otherwise not fulfill its contractual obligations. It is the latter category that is widely used in conjunction with construction projects.

Surety bonds differ from insurance protection in that they bind the surety (bonding company) to make good any loss of the obligee (owner, in most cases) caused by the principal's, or contractor's, failure to perform. There are four types of surety bonds used frequently to protect project owners:

1. *Bid Bonds.* The surety guarantees that if the named contractor's bid is accepted, the contractor will indeed sign the contract and secure any performance or payment bonds required. A bid bond ensures not contract performance but only contract acceptance.

2. *Performance Bonds.* These protect the owner, up to a specified bond penalty amount, against the contractor's failure to perform its contractual obligations. They are typically secured at time of contract award.

3. *Payment Bonds.* These protect the owner against loss should the contractor fail to pay its subcontractors or suppliers. They are designed to protect against mechanics' or materialmen's liens against the owner's property.

4. *Maintenance Bonds.* These bonds bind the surety (and contractor) to correct defects in the construction work that appear beyond contract closeout up to a specified time limit. An example would be a 10-year bond against leaks in a roofing system.

Most construction contracts awarded by the U.S. government are required to be covered by performance and payment bonds (the Miller Act). Most states have adopted "little Miller Acts" that require these bonds for state projects. Where these bonds are not required by law, project owners are free to choose whether to require bonds of their contractors.

## INSURANCE

No owner would knowingly tolerate the presence of an uninsured contractor operating on its project. The risks inherent in construction activity are too great for the contractor and the owner to omit adequate insurance coverage from the contract requirements. Most of us carry all types of insurance and are aware of the complexities of policies regarding medical, life, homeowners, and automobile insurance. Contractors carry insurance on their business and operations, and understanding all of the intracacies of construction insurance is a difficult task beyond the scope of this chapter. It is beneficial, however, to understand the basic insurance coverages applying to the construction industry and common ways that project owners deal with the protection of their needs through insurance.

Virtually all construction contracts require contractors to maintain minimum levels of coverage in certain areas. There are two principal reasons for this:

1. Should uninsured liabilities arising from an accident or occurrence affect the financial position of a contractor working on the project, the project could be in jeopardy or delayed.
2. The owner could be named as defendant or held responsible for damage arising out of the conduct of contractor operations.

As evidence of coverage, owners require contractors to submit "certificates of insurance" signed by a representative of the insuror and certifying that minimum insurance coverages, as required by the contract documents, are in effect. The insurance industry has adopted a standard form for such a certificate—an "accord" form.

Unless the nature of work a contractor will perform has particular hazards requiring special coverage (such as aircraft operations, or construction on or near bodies of water), the following four types of coverage are generally required:

1.  Comprehensive general liability
    a.  Operations–premises liability
    b.  Independent contractor's protective liability or owner's protective liability
    c.  Completed operations and products liability
    d.  Contractual liability
2.  Comprehensive automobile liability
3.  Workmen's compensation and employees' liability
4.  Umbrella excess liability

## LANGUAGE CONSIDERATIONS

Nothing is more tedious and frustrating than wading through a morass of obfuscating contract language. Construction contracts and RFPs are intended to transmit precise information to people who must act on it—and that action results in concrete (literally!) products that are extremely difficult to modify. Contract and specification language should be clear, concise, and direct.

The following guidelines are given in lieu of a detailed treatise on contract writing:

1.  Avoid "legalese." Unless they are absolutely needed for clarity, omit words such as *hereinafter, said* (as in "said party shall . . . "), *aforementioned*, and *herewith*. Avoid pompous phraseology, such as "the party of the first part, at which time and upon said notification, as pertains to the rights, duties, and obligations of one engaged in such an endeavor."

2.  The use of titles or terms interchangeably should be avoided. Although one frequently sees *owner, company, buyer*, and *company's representative*, or *contractor, vendor, seller*, and *supplier* used to refer to the owner and contractor, respectively, this must be purposely avoided. Throughout the documents, one and only one term should be used to identify a distinct party or item. Similarly, the terms *drawings, plans, contractor's drawings, shop drawings*, and so on, should not be used to identify the same thing. Each term has a defined, contractual meaning and should be used accordingly.

3.  Avoid the common temptation to repeat requirements. Say it once, say it where it should be said, and forget it. When the same requirement is stated in several places among the documents, besides being a nuisance to the reader, changing that requirement—finding

it, making sure you found each reference to it, and changing each one—invites unnecessary risk and effort.

4. Use each document for its intended purpose. Do not put technical requirements in the general conditions or commercial terms in the technical specifications or on the drawings.

5. Review and update standard or reference clauses and documents periodically to reflect changing needs, legal interpretations, governmental requirements, industry practices, and organizational preferences. Do not use 20-year-old documents even though "they seemed to work well last time."

6. Anticipate problems, misinterpretations, and scope changes and provide for these in the contract documents.

7. Get it in the contract! Contractors cannot be expected to read minds or anticipate and provide for particular owner requirements. If you want something, state it in the RFP and contract documents.

8. Consider use of the word *shall* to indicate required action or results of the contractor. Use the word *will* when describing activity by the owner or others. This helps clarify scope and responsibility assignments.

Additional suggestions are:

1. Consider standardization of technical specifications for a project, or, ideally, for all projects.

2. Do not ask for unusually complex or unnecessary bidder submittals (schedules, résumés, equipment lists, and so on) where they are not required for bid evaluation or contract administration.

3. Do not attach "reference" or "supplementary" specifications without defining which portions of such are applicable to the contractor's scope of work.

4. Number each paragraph of the commercial and technical documents for ease of reference during the bid cycle and contraction administration. Number pages as well.

5. Avoid excessive use of drawing notes in lieu of requirements in the specifications. Anyone who has been involved in revisions to bidding and contract documents will attest that drawings are much more difficult to revise and reissue than is printed material. Also, confusion and error may result when drawings containing contract-specific notes (such as "by owner," "by others," or "by C–15 contractor") are issued to contractors or parties other than those for whom the notes were intended.

## REPRESENTATIVE CASES

1.   A time-and-materials contract was bid for surveying services for an experimental nuclear reactor project. Bidders were requested to furnish hourly rates for standard crews including equipment, transportation, labor, material, and all overhead, profit, and other fees. For the eight-man crew, the bids were:

Bidder A: $240 per hour
Bidder B: $210 per hour
Bidder C: $232 per hour
Bidder D: $245 per hour

Since Bidder B was obviously low and all were well qualified, a contract was prepared referencing Bidder B's proposal. At the first progress payment meeting, contractor B presented an invoice as follows: 160 hours at $294/hr. = $47,040. The owner's contract administrator took immediate exception to the hourly charge, noting the contractor's proposal stated $210 per hour, not $292 per hour. The contractor then took a copy of the contract and showed in three places where his proposal (in fine print) clearly stated "a 40% markup is to be charged to quoted hourly rates to cover general and administrative costs." Further, his proposal stated that "fees will be determined in accordance with our standard procedure," which, as described in the fine print, included the 40% markup in making the hourly fee $210 × 1.4, or $294 per hour.

2.   For a highway construction project separate contracts were to be let for roadway lighting. Since the lighting contractor was originally not to install concrete standard bases (pole foundations), all drawings showing these were marked "not by electrical contractor" with a hand-drawn circle around each base. At the prebid conference, most bidders expressed a cost savings to the owner if bases were included in the standard installation. An addendum was issued to include the base installation as part of the scope of work. Electrical draftsmen, eager to make the change, spent five days removing the circles and notes from 188 drawings. When they had finished, the project owner reversed itself and ordered the bases removed from the scope of work again.

3.   When preparing piping diagrams and drawings, the mechanical engineer noted on each affected drawing that pipe stands were "by others"—thinking that the mechanical contractor would not be responsible for stands since this work was for the civil contractor. When the civil contractor was later given the same drawings, he claimed the

work was not included in his bid since the stands were clearly marked as "by others" on his contract drawings.

4.  Concerned that a test fill operation for the compaction of soil at a critical foundation be highly controlled, the specification writer laced the technical requirements with phrases such as:

> "The work will be under the constant supervision of the owner's representatives"

> "The owner will supervise and direct the test fill program to ensure satisfactory results"

At the same time, the contract administrator was preparing general and special conditions disclaiming any owner responsibility for supervision in order to maintain the independent contractor status of the successful bidder. While the RFP was on the street, a large personal injury lawsuit was concluded concerning the owner's previous project. The court denied liability on the owner's part from the injured worker's claim by upholding the independent contractor relationship and citing the absence of the right by the owner to supervise the contractor's work. In his remarks, the trial judge allowed that had the owner directly supervised the contractor's work on the previous project, he could be held responsible for the worker's safety and therefore liable for his injury.

5.  At a contract closeout party (after he had received and cashed his final check), the contractor's job superintendent allowed that he brought the lump-sum job in for his boss at $200,000 below the estimate. Later he divulged his technique in a spirit of greater conviviality: standard specs calling for hot-dipped galvanized piling were included in the RFP and the additional cost was included in his company's bid. After receiving the award and before ordering the piling, he found the galvanizing was not required—it was simply a standard specification clause the owner forgot to delete. He then left the party in his new luxury automobile, claiming that the owner left at least $150,000 on the bid table.

# CHAPTER 6

# BIDDER QUALIFICATION AND SELECTION

There are two principal methods by which owners inform potential bidders that they wish to receive bids: invitation and advertisement. Each constitutes notification by the owner that a future contract is to be awarded. Many publicly funded projects require that all potential bidders be notified and allowed to submit bids. This is commonly done under precise government or agency regulations, with an announcement of the upcoming bidding process publicized in newspapers, trade journals, or other media. As such, the initial contact between the owner and potential bidders is gained through "advertisement." In the public sector, however, bidders are usually notified on a selective basis; that is, only those contractors from whom the owner wishes a bid are notified.

Given the choice between this invitation process and the advertisement method, most owners would choose the former—they would prefer to select potential bidders rather than to leave the bidding process open to the public at large. This chapter contrasts control methods applicable to bidding when performed under advertised or invited purposes, discusses the bidder qualification effort encountered with the invitation process, and presents recommendations for owners when requesting bids by invitation.

## INVITATION VERSUS ADVERTISEMENT

Because advertising for bids is normally required by public projects, the government or agency for which construction is performed (or under whose jurisdiction bidding takes place) usually prescribes a specific bidding and bid evaluation procedure. All bids are normally welcomed, and the opening of bids is commonly done in public, with the content and amount of each bid becoming a matter of public record.

Because the process is dictated by regulation or statute, the persons responsible for soliciting bids and for eventually evaluating them are generally restricted as to what may or may not be done during the bidding period. For owners in this position, the process is simply one of following the rules established by others. And because the project usually involves the expenditure of public funds for the public good, great emphasis is placed on the price of submitted bids, often at the expense of other characteristics of the bids and the qualifications of the bidders themselves. As such, price becomes the governing criterion in selecting a successful bidder. Because bids will be received from virtually any source, the ensuing evaluation is often time-consuming and difficult. It is also difficult to control the quality of the proposals themselves as well as the quality of those who have submitted them.

When bidding is done by invitation, however, only those bidders prechosen by the owner are allowed to submit proposals for consideration. In order to determine from whom bids will be allowed, the owner must go through a qualification process—a detailed review of the capabilities and quality of potential bidders before choosing those who will in fact be asked to submit bids. And because bidding is done privately, confidentiality becomes an objective of all involved. This confidentiality concerns the identity of those who are allowed to bid as well as the results of the bid evaluation itself. Because most private owners who employ the invitation method are totally free as to whom they may award a contract, the low bidder (in terms of price) need not always be given the award or even considered for the award.

Other factors of interest to the owner enter into the award decision; factors such as financial stability, performance on previous contracts, adequacy of management, supervisory and craft personnel, and ability to control performance are often major considerations. Because the owner is in charge of bidding conditions and free to award to whomever it chooses, the bidding and evaluation processes are generally much more easy to control than if conducted under advertised conditions.

## THE QUALIFICATION PROCESS

Owners soliciting bids through invitation generally go about a structured process of determining the identity of contractors from whom a bid will be requested. This is the bidder qualification or prequalification effort leading to the establishment of a contract bidder list. There are several reasons for qualifying potential bidders before allowing them to bid:

1. *To Identify Stable Contractors.* Few things would be worse than allowing an unstable contractor to perform contract work. Whether it is financially unstable, has a history of poor performance, or has demonstrated a penchant for disputes and claims, it is much better to discover this beforehand and prevent this contractor from bidding than to deal with it as a project contractor. The major purpose of prebid qualification is to eliminate such contractors from consideration early in the formation cycle.

2. *To Prevent Contract Performance Problems.* Should a contractor be unsuitable for the work in question (even if it is stable and has a good overall performance reputation), it should not be allowed to bid. It may not have the resources to perform the particular job (men, equipment, funding) or have experience in contract-specific work scope or methods.

3. *To Ease the Evaluation Process.* Bidder qualification eliminates the need to spend time and money evaluating bids submitted by contractors who would never be considered for award, despite the contract price included in their bids. Those responsible for evaluating bids may then concentrate on the offerings, as they differ, made by qualified contractors.

4. *To Obtain the Positive Benefits of Contractor Competition.* The previous reasons given for bidder qualification may seem to indicate that the process always leads to the reduction in number of potential bidders. On the contrary, bidder qualification should be performed with the opposite goal in mind—to expand the list to the number of bidders needed to ensure receipt of competitive bids. If owners exhibit a major fault involving bidder qualification this may be it—the failure to seek the largest number of qualified bidders possible.

## QUALIFICATION PROCEDURE

A sound qualification process begins with the identification of a large number of potential bidders and the creation of a source list containing their names. This list may draw upon published registers of available contractors, of which there are many available. A preliminary bidder list is then made from the larger source list, and the included contractors are reviewed as to their quality standards, past performance, financial standing, and other factors deemed important to the owner.

The owner or its representatives then contact prospective bidders to determine their interest in the proposed work and intention to bid if requested. References are contacted to determine if these bidder candidates are financially and technically qualified. These references usually include banks, sureties (bonding companies), and previous owners or general contractors for whom the candidate has performed. To facilitate this process, many organizations maintain continuously updated bidder files. They may also prepare and issue standard questionnaires to contractors from time to time in order to solicit specific information that could bear on their qualification (such as financial statements, union affiliations, construction equipment inventory, geographical territory, and size of jobs they can undertake). These files can be manual or automated—with contractors indexed and sorted according to numerous factors. Standard, preprinted reference forms are also helpful in documenting telephone responses from selected references. Often references hesitate to disclose confidential information, but a call to the bidder candidate (and his or her subsequent calls) usually jars this information from a recalcitrant source.

A common error made by owners is to allow bidder questionnaires, reference documentation, or even bidder lists themselves to languish for months or years without being updated or reviewed to include any changes that have occurred. In order for any qualification to be effective, it should be based upon recent information concerning the contractors in question. Procedures should call for periodic revision of bidder qualification files and bidder lists.

Another problem concerns the rigidity with which company bidder lists are often used. Most companies who attempt to establish a company-wide bidder list—covering many different scopes of work, geographical locations, price limitations, and the like—find that their overall list is so restrictive that it is virtually impossible to find competitive bidders in any number for a specific job. And it is entirely possible that any one bidder may be qualified to perform certain work yet totally unqualified to perform others. For these reasons, it is recommended that general, all-encompassing bidder lists be avoided. In their absence, company-wide source lists may be used, with potential contractors taken from these lists and further qualified according to the specific requirements of the job in question.

## INTERNAL SOURCES OF BIDDER INFORMATION

One valuable source of information is often overlooked by owners and their project or construction management consultants. This is the performance by the contractor on previous or concurrent projects for them. For this reason, it is recommended that periodic performance reports be issued for each contractor while it is under contract and that a simple performance appraisal (see Exhibit 21) be completed and sent to the bidder files when each contract is closed out. It is surprising that in our haste to find out about a contractor, we often neglect internal sources of information.

## THE TIMING OF QUALIFICATION

The degree of qualification required before allowing a contractor to bid is a subjective determination. The same is true for the proper number of bidders needed to ensure a representative sample of prices and the lowest available cost. Arguments can be made for a perfunctory (or no) qualification prior to bidding since all but one of the bidders will be rejected—that only the apparent successful bidder need be investigated. Problems with this approach are:

1.  Time may not permit adequate qualification once bids are received.
2.  If bidders are obviously unqualified, why waste time and effort receiving and evaluating their bids?
3.  Compromised standards of quality may be more readily adapted in view of low bids (price may unduly influence judgment).
4.  The owner may discover that *no* bidders are qualified.
5.  It could be embarrassing or require unnecessary documentation, approvals, and justification to award the contract to someone other than the lowest bidders simply because they are not qualified.
6.  Postponement of bidder qualification until time of bid evaluation entails all of the problems associated with advertisement.

The rule of reason should apply to this dilemma. If qualification is relatively simple—for example, low-risk work involved, or most bidders are well known to the owner—or the qualification process is of short duration, qualify prior to bidding and avoid problems later. If qualification standards are exceptionally high and selective, or a time-consuming and expensive survey is required, make a preliminary screening prior to bidding and conduct detailed qualification as part of the bid evaluation process, focusing on only those candidate(s) who appear to be successful. The needs of the project, particulars of the contract, and schedule considerations all enter into this decision. As far as confidentiality goes, it is suggested that each qualified bidder (or others who inquire) be told only whether or not his company is on the bidder list.

## REPRESENTATIVE CASES

1.  Mr. X, an executive of an A–E firm's Western Division and project manager for a hydroelectric dam project, had just signed a contract award recommendation to let the general construction contract. By now the letter was in the mail advising his firm's client, the project owner, to choose Contractor A for the work. Although Contractor A was not the low bidder, his contract people had chosen them after an extensive analysis of the bids. Mr. X phoned his golf partner, corporate attorney for the A–E firm, and was told by the lawyer's secretary that he was in Chicago taking depositions on "the Contractor A Case"! After several frantic phone calls, Mr. X finally ascertained that Contractor A and the A–E firm had been entwined in a bloody legal battle for years—with Contractor A brutalizing them in the courts and press over a previous dam project for their eastern division. Mr. X was last seen chasing the mailman as he left the A–E firm's headquarters.

2.  The contract administrator for a grass-roots manufacturing plant mailed RFPs to five contractors for the interior architectural work. All five had bid on the previous project some eight years ago and time was short, so he allowed only two weeks for bidding. On the bid due date, no bids were received. The contract administrator called all five bidders and received the following responses:

Bidder #1: Has since gone out of business

Bidder #2: Owner and general manager "out of the office" (spending four years in the federal penitentiary for conspiracy and fraud)

Bidder #3: Acquired two years ago by a huge conglomerate, which decided to drop the contracting "product line"

Bidder #4: Decided not to renew contractor's license in that state due to infrequent business there

Bidder #5: Decided not to bid "partially because we are in receivership and partially because you threw us off your last project"

3.  In a great hurry to qualify additional bidders, a contract administrator cannot get past the lowest clerk who answers the telephone at a contractor's bank. She refuses to divulge "confidential information" concerning Contractor A's financial status. Frustrated and pressured, the administrator calls the chief executive officer of Contractor A and tells him that he will not be asked to bid unless this information is provided by the bank. The administrator finishes the call and walks across the office for a cup of coffee. When he returns there are three urgent phone message slips from the bank president. He returns the call, during which he is offered a complete set of financial statements on Contractor A and invited to dinner at the bank president's private club to receive them.

# CHAPTER 7

# ISSUING REQUESTS
# FOR PROPOSALS

Once qualified bidders are selected for a particular contract, and the technical specifications, drawings, and other bid information have been prepared, the contract administrator is able to prepare a request for proposal (bid solicitation) and transmit it to the prospective bidders. Requests for proposals (RFPs) are composed of the following documents:

Invitation (instruction to bidders)
Proposal forms
Agreement
General conditions
Special conditions
Technical specifications
Drawings

During the initial contract document preparation period for the project (see Chapter 5), the format for all these documents is established, and certain ones are not altered for each RFP (such as the general conditions). The RFP development process, then, entails the completion of the remaining documents with variable information pertaining to the particular contract at hand. Each of these documents is discussed in the following sections.

The exhibits for this chapter appear on pages 107–126.

## BIDDING DOCUMENTS

### INVITATION

Since this letter informs the bidders of the requirements for submitting bids, particular information is included, such as the bid due date and time, number of copies of bids expected, project name and location, name and identification of contract, and date of prebid meeting, site visit, and so on. A sample letter is shown as Exhibit 2. It is helpful to alert bidders to any unusual bid requirements in this letter and to ask them to acknowledge receipt of the RFP as well as to return notification of their intention to bid (or not) to the sender. Some owners include a preaddressed postcard for this purpose. In order to identify the returning bids, preaddressed mailing envelopes or address labels can also be included.

### PROPOSAL FORM

This document is prepared by the owner and completed by the bidders. The intention of a well-defined and well-structured RFP is that the proposal form is the only document altered (completed) by bidders. The proposal should be designed parallel to the agreement so that bidder-submitted data can easily be transposed to the agreement once an award is made.

Since the proposal form is to contain the bidders' prices, its pricing section should be carefully prepared to solicit the intended contractual strategy and pricing alternatives. This usually constitutes the greatest effort of the proposal form preparation: designing the pricing section. It will be affected by the type of pricing selected (lump sum, unit prices, cost plus, and so on) and should, in all cases, provide for flexible pricing of any envisioned changes.

A sample proposal form for a lump-sum contract is included as Exhibit 3. Note that it provides for preestablished unit prices for additions and deletions and requires the bidder to enter proposed markups for any cost-plus changes. Should the base contract be cost reimbursable, the form would be markedly different, as it would identify the allowable costs in great detail. This information is not usually required for a lump-sum contract.

### Structuring Pricing Data

In soliciting pricing information, the natural tendency is to ask for only that which is necessary to select a low bidder. In the case of a lump-sum contract, for example, this may be only one dollar amount. Asking the bidders to show the components of this price (broken down by measurable, objective increments, in addition to the gross amount) will be helpful for a number of reasons:

1.  Measurement of performance and determination of progress payments will be based on an objective determination of actual accomplishment rather than on subjective percentages of the total (progress billings and payments are discussed further in Chapter 12).
2.  Unbalanced bids, such as front-end loading, will be surfaced.
3.  Bidder error or misunderstanding as to scope of work will be isolated.
4.  Baseline costs for additional or deleted segments of work will be established.
5.  Historical, real-world cost data can be used to refine detailed owner project estimates for this and other projects.

In many cases, conflicts arise when the cost data is structured by one of the project management participants without regard for the needs of the other. Here are some typical situations:

Project cost estimates are prepared at a conceptual level or by element of cost orientation such that it is difficult to reconcile them with actual contract or subcontract line items.

Pricing structures in the RFP issued by the contract administrator do not request cost data from bidders in a manner that facilitates calculation of "life-cycle" costs by the cost engineer.

RFP price sheets cause the cost engineer to reestimate the contract for the fair price estimate (fair price estimates are discussed in Chapter 9).

Contract line items are structured to match the way the project was estimated rather than the way the work will be performed. This makes it difficult to determine the value of work completed for progress payments.

A certain amount of give and take will be required, and the process of cost-related input to the bidding documents is usually an iterative exercise. But when it is successfully accomplished, all participants should be satisfied. The contract administrator will have bids and eventual contract pricing items structured in a way that facilitates bid evaluation and contract administration, whereas the cost engineer will be better able to update his or her estimating data base and report actual and projected costs throughout the project.

## ADDITIONAL SUBMITTALS

The proposal form should also list the required additional submittals and provide for the bidder's signature. It then becomes, in effect, an offer on the part of the bidder to perform the work, as defined, for the given price. Once accepted by the owner, a contract is formed.

When different options are solicited, or bidders are asked to modify or further define technical requirements, care should be taken to specify these explicitly in the RFP—and provisions should be made in the proposal form for each base bid, option, alternate, and details of bidder-provided technical or nonprice information. For example, if a performance specification is used, bidders can be encouraged to propose any number of alternatives that meet the standards given. In order to evaluate these alternatives, the proposal form should be explicit and easy to understand. When technical submittals are needed to evaluate these properly, their format, content, and specific data needed should be given.

It is advisable to include several extra copies of the proposal form with the RFP (say, 5). This is convenient for bidders but also allows the owner to ask for more than one completed proposal to ease its evaluation task. If several internal groups are to evaluate bids simultaneously (for example, cost, technical, commercial), having an adequate number of forms facilitates this process for the owner. Sometimes owners specify that pricing information be contained on some copies and omitted from others. This allows noncost evaluations to be made without stringent bid security measures.

## SPECIMEN CONTRACT DOCUMENTS

The documents described in the following subsections are included with an RFP as specimens of the anticipated contract documents.

## AGREEMENT

This document, along with the remainder of the package, is sent to inform the bidders of the contents of the final contract once a bid is accepted. Upon award, it is completed and signed and forms the heart of the contract documents. A sample is shown as Exhibit 4.

## GENERAL CONDITIONS

These should require no specific alteration for the RFP. A sample table of contents for a set of general conditions is given in Exhibit 5.

## SPECIAL CONDITIONS

Contract-specific commercial terms are included here. As shown in Exhibit 6, the special conditions describe the scope of work, requirements for contractor's employees, material, equipment and labor to be furnished by the contractor, owner, and others, schedule and general sequence of work required; list the applicable drawings; and define how work will be measured and payment made.

## TECHNICAL SPECIFICATIONS

Before the RFP is prepared, the technical specifications (specs) must be reviewed to ensure that they are complete, that the scope of work descriptions contained in the other documents agree with the specs, and that no commercial or legal terms or conditions are included in them.

## DRAWINGS

Usually at least one copy of each drawing required to perform the work is included in the RFP packages. Some owners include two or more copies for the bidders' convenience. The drawings should be listed as to title, number, revision status, and date of issue in the RFP (as suggested here in the special conditions) so the bidders can check to ensure that they have all proper versions of the drawings. It is not unusual for the wrong drawing to be mailed to bidders or for some drawings to be omitted, illegible, or outdated. The risk of error is intensified under schedule pressures and with large quantities of drawings and fast-tracked engineering effort.

## ISSUING THE REQUEST FOR PROPOSAL

Once prepared and reviewed by all affected parties, (engineering, estimating, project management), the RFPs are ready to be mailed or submitted to potential bidders. Written documents are usually bound in a booklet or binder; drawings are printed and mailed. Several copies are sent to each bidder to facilitate eventual transmittal of the documents to subbidders. Some owners issue reproducible drawings, such as sepia copies, to bidders for this reason. It is recommended that the contract administrator contact each bidder within a few days to ensure that it has received the RFP. When fair price estimates are used (see chapter 9), the estimator or cost engineer should receive the identical package as sent to bidders and be allowed the identical time to complete his "bid." Once the RFPs are issued ("on the street"), requests may come to the owner for additional copies of documents or drawings. It is up to the owner to set a policy on reproduction and mailing costs. Often, these requests come from subbidders who will be quoting a portion of the contract work to one or more eventual bidders. A good control policy is to respond only to *bidder* requests, not to subbidder needs. If subbidders need copies (or information), they should be instructed to communicate through bidders. This brings up the question of whether subbidders should be told who the prime bidders are. It is an owner's prerogative to do so when bidding is by invitation, but in order to avoid problems, it is suggested that this not be done. One exception occurs when specific products or suppliers are required by name in the RFP. These suppliers (subbidders) should be told from whom to expect inquiries.

Some owners require all RFP documents to be returned with the bid, or otherwise if the bidder declines to bid. Other owners will establish RFP "viewing rooms" at various locations where interested subbidders may view or copy the RFP documents. This is unusual in invitation-only circumstances.

## REPRESENTATIVE CASES

1.  A contractor had been selected by the owner of a proposed $200 million office tower complex. The design was fairly well defined, but recent fluctuations in the market price of major building materials in addition to the known penchant of the owner's president to change his mind concerning interior design led the owner to negotiate a guaranteed maximum price with the potential contractor. Since only one contract was to be awarded for the project, the task of designing the contract and bidding documents and that of preparing a specific RFP were combined. The intention was for the contractor to "bid" the guaranteed maximum amount using the RFP documents.

Unfortunately, in designing the proposal form, no thought was given to the eventual task of evaluating the contractor's guaranteed maximum amount. When the bid was received, the owner found that the contractor had broken his bid amount down according to the project's major structures: office tower, service building, parking garage, pedestrian mall shops, and so on. The project estimate, however, had been prepared by the architectural consultant according to material quantities and elements of costs: structural steel, concrete, elevators, labor, concrete formwork, roofing, HVAC systems, and so on. Needless to say, meaningful comparisons between the two price structures were impossible. And, as luck would have it, the bottom-line guaranteed maximum price exceeded the architect's estimate by 40%. The owner, under schedule pressure, was forced to either:

1.  Accept the contractor's bid and proceed on faith that it was reasonable.
2.  Decrease the size or quality of the project to meet the budget established as a result of the architect's estimate. But where to cut?
3.  Reestimate the project according to the contractor's bid format. This would make the evaluation feasible, but the resulting contract price structure would make objective measurement of progress and corresponding progress payments difficult.
4.  Rebid the project after defining an objective work breakdown structure, reestimating the project according to this structure, and asking the contractor to conform his bid to such a structure.

For the benefit of all concerned, particularly its own, the owner chose the proper course of action: option 4.

2.  A construction management firm decided that since bid drawings change so frequently during the bid cycle, the use of a drawing list in the special conditions would only add another section needing revisions through addenda. A lump-sum contract served as the basis for reversing this policy. Of 362 drawings issued for bid, with revisions occurring daily, it was inevitable that a contractor would receive and base his bid on some superseded drawings. In this case, the low bidder and eventual contractor did not include in his bid 19 large electrical manholes, the substitution of nickel-chromium pipe for carbon steel pipe, the addition of concrete pads for buried storage tanks, and several other expensive changes brought about through drawing revisions as the RFP was being prepared. Through human error, all other bidders received and noted the drawing changes and were therefore higher in bidding. Resulting change orders to the eventual contractor had the cumulative effect of making him what would have been the highest bidder. Had he received the appropriate bid drawings, this problem would have been avoided.

3.  General conditions used for cost-plus contracts issued by a consulting firm called for reimbursement of "all reasonable overhead, general and administrative expenses providing such are substantiated by purchase or rental invoices, payroll records, stock depletion records, and the like." This broad and general language was the cause of delight and subsequent abuse for a general contractor hired to design and build an industrial plant for the consultant's client. Absent precise conditions specifying allowable and allocatable expenses for the project, the general contractor passed through expenses that normally would be denied. Only a small sampling of such expenses is given here to demonstrate the need for precise definition of reimbursable expenses in every cost-plus contract:

New office computer

New radios and transmitters

New vehicles for all management personnel

New ambulance

New drafting equipment

Lease or purchase of computer-assisted drafting hardware

Complete overhauls of all cranes, trucks, compressors, and other major equipment

Purchase of 30 welding machines

Tuition and expense reimbursement for attendance at 27 seminars by project personnel

Purchase of two copy machines and one word-processing computer

New hard hats and safety equipment

One-year personal services consulting contract for a retired member of the contractor's board of directors

Travel expenses for project manager and contractor's president to Florida, Hawaii, Bermuda, and London to meet vendors and/or visit similar projects under construction

And so on. . . .

**EXHIBIT 2. Sample Invitation to Bid**

Date

(Bidder's name and address)

Gentlemen:

(Owner's name, project name, description of work, and contract number)

You are hereby invited to submit your proposal for subject work located (site location) for (owner's name).

Proposals are due before (bid due date, time, and time zone) in the offices of _____ at the following address:

_____

_____

_____

ATTN: _____

Preparation of proposals shall be at the Bidder's expense, in accordance with the enclosed Specimen Contract Documents, and on the Proposal Forms provided herewith. Interlineations or alterations to these forms will not be acceptable. Bidders shall list any exceptions or clarifications on a separate sheet of paper, bearing their letterhead, with suitable references to the specific items as addressed in the Contract Documents.

Proposals shall be submitted in the following manner.

1 signed original (priced)
3 signed copies (priced)
3 unsigned copies (complete except for price listing)

No proposal security is required.

Requests for interpretation of the Request for Proposal documents shall be submitted in writing to Mr. _____ at the above address. All changes to these documents will be made by the Owner in the form of written addenda issued to all Bidders.

Bidders are advised (or required for consideration of their proposals) to visit the job site to acquaint themselves thoroughly with all conditions which may affect their proposals or performance of the work. Arrangements for site visits shall be made with the Owner through Mr. _____ at _____ .

*(continued on p. 108)*

**EXHIBIT 2.** *(continued)*

A prebid meeting will be held on (date) at (time) at (location). Failure of a Bidder to attend may be cause for rejection of his proposal.

Should additional copies of the Request for Proposal documents be required, contact Mr. _____ . Requests from noninvited contractors, subcontractors, or other third parties will not be honored.

Bidders are hereby informed that the contract between the Owner and the successful Bidder will be the legally controlling document and that all communications, verbal or written, made prior to or in addition to the Contract Documents signed by both parties will be abrogated, withdrawn, and of no legal force or effect.

The Owner reserves the right to open Proposals privately and unannounced. The Owner does not obligate itself to accept the lowest or any other bid and reserves the right to reject any and all bids.

Kindly inform Mr. _____ by (date) as to your intention to bid and to attend the scheduled prebid meeting.

Very truly yours,

_____
(signature)

**EXHIBIT 3.  Sample Proposal**

PROPOSAL
FOR
(Description of work to be covered
by the contract and contract number)
FOR
(Project description)
OF
(Owner's name and location)

The undersigned Bidder herewith proposes to do all the work, perform all the services, and furnish all the materials, except the specific materials and services to be furnished by the Owner, required for (description of work to be covered by the contract) for the (project description) of (owner's name); the work site being located at (location of work). The scope of work is set forth in the attached specimen Contract Documents labeled Contract (contract number), and all addenda which may be issued thereto.

The Bidder agrees that in case the Contract is awarded to him, he will begin actual work on _____ or within _____ calendar days after receipt of notice to proceed. The Bidder has attached to this Proposal a schedule for the work and agrees to maintain a rate of progress set forth in said schedule and that he will complete the work on or before the dates of completion required by Contract Document (contract number).

The Bidder proposes to complete all work as defined in Contract Document (contract number) for the lump-sum amount of _____ dollars ($_____), which is composed of _____ dollars ($_____) material cost and _____ dollars ($_____) labor cost.

A breakdown of the Contract Price is as follows:

## I. Lump Sum

| Item no. | Description | Material | Labor |
|---|---|---|---|
| 1 | | $ | $ |
| 2 | | $ | $ |
| 3 | | $ | $ |
| | Total material cost | $ | |
| | Total labor cost | | $ |
| | Total lump-sum cost | $ | |

## II. Additions and Deletions

The following unit prices will apply for additions or deletions to the amounts included under Item I above:

| | | | Unit price | |
|---|---|---|---|---|
| Item no. | Description | Unit of measurement | Material | Labor |
| 1 | | | $ | $ |
| 2 | | | $ | $ |
| 3 | | | $ | $ |

## III. Unit Prices (if used)

The following unit prices will apply to work not included under Item I above:

*(continued on p. 110)*

**EXHIBIT 3.** *(continued)*

| Item no. | Description | Estimated quantity | Unit of measurement | Unit price | |
|----------|-------------|--------------------|---------------------|------------|---|
| | | | | Material | Labor |
| 1 | | | | $ | $ |
| 2 | | | | $ | $ |
| 3 | | | | $ | $ |
| 4 | | | | $ | $ |

Should the Bidder, as Contractor, be requested to perform any additional work on a cost-plus basis, the proposed fees, as defined in Contract Document (number), General Conditions, Article (changes clause), are as follows:

_____ % for direct labor

_____ % for material

_____ % for subcontracted work

When performing cost-plus work, craft labor wages including fringes and benefits shall be in accordance with prevailing local rates. The rates shall be submitted in compliance with the attached Sample Wage Rate Breakdown to be completed by the Bidder and submitted with his Bid.

Equipment rates for additional work shall be discounted at _____ % of those shown in the current A.E.D. Rental List or as shown on the attached Equipment Rental Schedule to be completed by the Bidder and submitted with his Bid. Rental lists or schedules shall show rates for monthly, weekly, and daily rentals, and those rates producing the most economical result for the Owner shall apply, in accordance with the actual circumstances of the case in each instance.

The Bidder proposes to credit the Owner the following amounts to be deducted from the total Contract Price, in the event the Owner provides the stated material or services:

1.  (List of material or services owner may provide          $(          )
2.  at its option, over and above those listed in
    the specimen contract documents)                          $(          )
3.                                                             $(          )

The Bidder requires the following temporary facilities:

1.  Site space requirements for field office and storage          _____ sq. ft.
2.  Electric power requirements                                   _____ kVA

The Bidder agrees that this Proposal constitutes a firm offer to the Owner and cannot be withdrawn for ninety (90) calendar days after the due date. The

Bidder also agrees that the prices listed in the Schedule of Prices are firm and shall remain so throughout the performance of the work.

The Bidder proposes to pay all federal, state and local income, gross receipts, and franchise taxes due and payable by Contractor in connection with all work provided for in the Contract Documents, to make any and all payroll deductions required by law, and to hold the Owner and the Engineer harmless from any liability on account of any such taxes or withholdings.

The Bidder agrees to perform any additional work as requested under the terms and conditions of the Contract Documents.

The Bidder proposes to subcontract specific portions of the work to the following subcontractors:

| Subcontract work | Subcontractor's name and address | Estimated % of total contract amount |
|---|---|---|
| | | |

No work other than that listed above shall be subcontracted unless approved by the Owner. Bidder agrees that, if he is awarded the Contract, he will submit to the Owner copies of all subcontracts and necessary insurance information, in accordance with the General Conditions, before each subcontractor starts work.

The Bidder shall attach the following information to his Proposal.

1.  Proposed job superintendent's résumé, including a listing of recent jobs of this nature, areas of responsibility, duration, and job references including names and telephone numbers
2.  A simplified Critical Path Method precedent diagram or a bar chart indicating Bidder's suggested start and completion dates for the principal sections of the work
3.  Manpower loading curves for the various crafts under his direction
4.  Equipment rental schedule
5.  Sample wage rate breakdown
6.  (List any other submittals required with the proposal, such as a list of foreign materials, organization chart with names of proposed field staff, proposed delivery schedules, and any submittals required with the bid by the technical specifications.)

The Bidder proposes to furnish to the Owner, if requested, within ten (10) days after execution of the Contract, a surety company bond, in the form provided herein, for the faithful performance of the Contract including the payment of

*(continued on p. 112)*

111

**EXHIBIT 3.** *(continued)*

all labor and indebtedness to equipment rental firms, material suppliers, and any other vendors or subcontractors to whom the Bidder may be obligated in respect to the performance of the Contract. The penalty of the performance bond shall be one-hundred percent (100%) of the Contract Price. Cost of such bond will be borne by the Owner and shall not be included in the prices listed in the Schedule of Work.

Proposed surety is:

Full Name _____

Address _____

A Corporation organized under the laws of the state of _____ and duly authorized to transact business in the state of _____ . Cost of premium to be borne by Owner is $ _____ per $1,000 of bond.

Any jurisdictional dispute that may arise in connection with work performed pursuant to the terms of the Contract shall be settled in accordance with the Plan for the Settlement of Jurisdictional Disputes in the Construction Industry and any decision by the Impartial Jurisdictional Disputes Board shall be final and binding.

In performing the work hereunder, the Bidder, as Contractor, proposes to adopt wages, working conditions, and other employment policies that meet with the approval of the Owner and, where applicable, shall comply with the wages and working conditions established in the agreements with local unions affiliated with the Building and Construction Trades Department (AFL-CIO) having jurisdiction over the specified work, provided that the Bidder, as Contractor, will not be required to violate any applicable federal, state, or local codes, regulations, or statutory provisions.

The undersigned Bidder declares that he or his agent has carefully examined the location of the proposed work, the Contract Documents, and the Drawings for the Contract, is informed as to local conditions affecting the proposed work, and has completed the requested attached information, and that he understands that the Owner reserves the right to reject any or all bids.

The Bidder affirms that, in making this Proposal, neither the Bidder, any company it represents, nor anyone on its or its company's behalf, has directly or indirectly entered into any combination, collusion, undertaking, or agreement with any other bidder or bidders to maintain the prices of said work or any compact to prevent any other bidder or bidders from bidding on the work described in this Proposal. The Bidder further affirms that this Proposal is made without regard or reference to any other bidder and that it is made without any agreement, understanding, or combination with any person or persons concerning such bidding in any way or manner whatsoever.

The Bidder is (please indicate below):

An Individual _____

A Corporation of the state of _____

A Partnership consisting of _____

A Joint Venture comprised of _____

Full company name _____

By _____

Title _____

Date _____

Business address _____

_____

Telephone number _____

(State) Contractor license number _____

The Bidder states that the following individuals are authorized signatories of contracts binding on the Bidder for not less than the total Contract Price as stated herein:

| Name | Title |
|---|---|
| _____ | _____ |
| _____ | _____ |
| _____ | _____ |

The Bidder acknowledges receipt and consideration of the following Addenda to the Request for Proposal:

| Addendum no. | Date issued |
|---|---|
| _____ | _____ |
| _____ | _____ |
| _____ | _____ |
| _____ | _____ |

**Sample Wage Rate Breakdown (to be completed by Bidder)**

Contractor: _____

Craft: _____

Classification: _____

Union local no. and address: _____

_____

*(continued on p. 114)*

EXHIBIT 3. *(continued)*

| Description | Straight time | Rate (per hour) premium time | Shift time |
|---|---|---|---|
| (1) Base wage | $ | $ | $ |
| (2) Fringes and benefits (list individually contributions to funds and other fringe benefits required under terms of collective bargaining agreements) | | | |
| a. | $ | $ | $ |
| b. | $ | $ | $ |
| c. | $ | $ | $ |
| d. | $ | $ | $ |
| e. | etc. | $ | $ |
| (3) Payroll taxes and unemployment insurance (list individually Social Security taxes, Unemployment Insurance, Workmen's Compensation, and so on, including amounts and percentages) | | | |
| a.        @      % | $ | $ | $ |
| b.        @      % | $ | $ | $ |
| c.        @      % | $ | $ | $ |
| d.        @      % | $ | $ | $ |
| e.        @      % | $ | $ | $ |
| (4) Markup for overhead and profit (per contract)           % | $ | Not applicable | |
| (5) Total wage rate (Total of items 1 through 4) per hour. | $ | $ | $ |

Note: Separate sheets shall be submitted for each classification (such as apprentice, journeyman, and foreman).

### Equipment Rental Schedule (to be completed by Bidder)

The following rates include fuel, lubrication, and all maintenance but without operator. No overtime premium will be paid on equipment. The Owner reserves the right to select the rental period that proves to it to be the most economical.

| Equipment | Description | Rate per hour | Rate per week | Rate per month |
|-----------|-------------|---------------|---------------|----------------|
|           |             |               |               |                |

**EXHIBIT 4. Sample Agreement**

AGREEMENT
FOR
(description of the work)
FOR
(Owner's name)
(project description)

This AGREEMENT, made and entered into (by letter of intent as of the, or) this _____ day of _____ in the year One Thousand Nine Hundred and _____(19___ ), by and between (owner's name) hereinafter called the Owner, and (contractor's name), whose office is located at (contractor's business address), hereinafter called the Contractor,

WITNESSETH that the parties to these presents, each in consideration of the undertakings, promises, and agreements on the part of the other herein contained, have undertaken, promised and agreed, and do hereby undertake, promise and agree, the Owner for itself, its successors and assigns, and the Contractor for himself, his successors and assigns, as follows:

FIRST—The Contractor agrees, at his own sole cost and expense, to perform all the labor and services, and to furnish all the materials and permanent equipment except as herein otherwise specifically set forth, and to furnish all the construction plant and equipment necessary to complete in a good, substantial, workmanlike and approved manner within the time hereinafter specified and in accordance with the terms, conditions and provisions of this Contract, and of the instructions, orders and directions of the Owner made in accordance with this Contract, the (description of the work), at the Owner's site located (project location).

*(continued on p. 116)*
**115**

**EXHIBIT 4.** *(continued)*

The above works are more completely described in the General Conditions, Special Conditions, Technical Specifications, and the Drawings which are part hereof.

The Contractor agrees to protect the said works from damage by the elements or otherwise until completed and delivered in accordance with the Contract Documents.

SECOND—The Owner agrees to furnish to the Contractor the materials and services as provided for under Article (number) of the Special Conditions.

THIRD—The Contractor agrees to comply with the schedule as provided for under Article (number) of the Special Conditions.

FOURTH—The Owner agrees to pay and the Contractor agrees to accept as full compensation, satisfaction, and discharge for all work performed and all materials installed, and for all costs and expenses incurred, damages sustained and for each and every matter, thing or act performed, furnished or suffered in the full and complete performance and completion of the work of this Contract in accordance with the terms, conditions and provisions thereof, and of the instructions, ordered and directions of the Engineer thereunder, except changes or additions covered by Contract Change Orders, which shall be paid for as provided in the General Conditions, a lump sum equal to _____ Dollars ($           ).

A Breakdown of the Contract Price is as follows:

## Lump-Sum Schedule

| | | Amount | |
|---|---|---|---|
| Item no. | Description | Material | Labor |
| 1 | (breakdown of lump-sum items, as shown in the schedule of prices, Item I of the proposal) | $ | $ |
| 2 | | $ | $ |
| 3 | | $ | $ |
| 4 | | $ | $ |
| 5 | | $ | $ |
| 6 | | $ | $ |
| | Total material price | $ | $ |
| | Total labor price | | $ |
| | Total contract price | $ | |

For additions and deletions to the work covered in the Lump-Sum Schedule above, the Contract Price will be measured and derived from the amount of actual work performed and materials installed, as determined by the Engineer under each item of the following schedule, multiplied by the unit price applicable to each such item as set forth in the following schedule:

### Unit Price Schedule for Additions and Deletions

| | | | Unit price | |
|---|---|---|---|---|
| Item no. | Description | Unit of measure | Material | Labor |
| 1 | (unit price schedule for additions and deletions as shown in Item II of the schedule of prices in the proposal) | | $ | $ |
| 2 | | | $ | $ |
| 3 | | | $ | $ |
| 4 | | | $ | $ |
| 5 | | | $ | $ |

For work that is included in the scope of work under this Contract, but that is not covered by the Lump-Sum Schedule above, the following Unit Price Schedule shall apply. Payment will be a sum to be measured and derived from the amount of actual work performed and materials installed, as determined by the Engineer, multiplied by the unit price applicable to each item as set forth in the following schedule:

### Unit Price Schedule

| | | | Unit price | |
|---|---|---|---|---|
| Item no. | Description | Unit of measure | Material | Labor |
| 1 | (unit price schedule as shown in Item III of the schedule of prices in the proposal. Note that estimated quantities are not given in the contract agreement but serve only as a basis for bidding and are therefore withdrawn after receipt of proposals.) | | $ | $ |
| 2 | | | $ | $ |
| 3 | | | $ | $ |

*(continued on p. 118)*

**EXHIBIT 4.** *(continued)*

The measurement and payment of all items included in the above schedules shall be as prescribed in Article (number) "Measurement and Payment" of the Special Conditions.

Should the Contractor be requested to perform any additional work for which he will be compensated in a cost-plus manner as specified in Article (number) of the General Conditions, the Contractor's fees for such work shall be as set forth in the following schedule. Equipment rental rates paid shall be as shown on the attached Equipment Rental Schedule, or shall be discounted at _____ % of the current A.E.D. Rental List if not shown on the attached Schedule. Equipment rental reimbursement will be on the basis of monthly, weekly, or daily rates as actually contracted for by the Contractor with the equipment lessor.

### Additional Work Pricing Schedule

(Insert the pricing schedule for additional work as shown in the proposal.)

FIFTH—All the provisions contained in this Agreement, General Conditions, Special Conditions, Technical Specifications, and Drawings and those items included by reference shall be deemed terms and conditions of the Contract as if fully embodied herein.

These Contract Documents are complementary, and what is called for by one of them shall be as binding as if called for by all.

SIXTH—Without limiting any of the other obligations or liabilities of the Contractor, the Contractor shall adhere to the General Conditions and provide and maintain insurance in accordance with the provisions set forth therein until the work is completed and accepted by the Owner.

The contractor shall before commencing work on this Contract, deliver to (name) at (address) two copies of Certificates of Insurance, completed by his insurance carrier or agent certifying that minimum insurance coverages required by the Contract Documents are in effect.

SEVENTH—The Contractor agrees to pay all federal, state, and local income, gross receipts and franchise taxes and taxes of every other nature due and payable by the Contractor in connection with all the work provided for in the Contract, and to make any and all payroll deductions required by law, and to hold the Owner and its agents harmless from any liability on account of any such taxes or withholdings.

(Insert specific provisions for local sales and/or use taxes as applicable.)

EIGHTH—Any jurisdictional dispute that may arise in connection with work performed pursuant to the terms of the Contract shall be settled in accordance with the Plan for the Settlement of Jurisdictional Disputes in the Construction Industry and any decision by the Impartial Jurisdictional Disputes Board shall be final and binding.

In performing the work hereunder, the Contractor shall adopt wages, working conditions and other employment policies that meet with the approval of the Owner, and, where applicable, shall comply with wages and working conditions established in the agreements with local unions affiliated with the Building and Construction Trades Department (AFL –CIO) having jurisdiction over the specified work provided that the Contractor will not be required to violate any applicable federal, state, or local codes, regulations, or statutory provisions. (Article Eight is applicable when union labor is required by a project agreement or other understandings and may not always be applicable.)

NINTH—The Contractor, within ten (10) days after the execution of this Contract, shall furnish to the Owner a surety company bond in the form hereto attached with such surety as shall be satisfactory to the Owner, conditioned in the amount of one hundred percent (100%) of the Contract Price for the faithful performance of this Contract.

(delete above article if bond is not required)

IN WITNESS WHEREOF, the Parties hereto have duly executed this Agreement the day and year first written.

 

_____ (owner's name)

By _____
                 Name

_____
                 Title

Witness: _____

 

_____ (contractor's name)

By _____
                 Name

_____
                 Title

Witness: _____

**EXHIBIT 5.  Sample General Conditions for Construction Contracts (Index Only)**

GENERAL CONDITIONS
FOR
(owner's name)
(project description)
INDEX

Page

**EXHIBIT 6.   Sample Special Conditions**

SPECIAL CONDITIONS
FOR
(description of work and contract number)
FOR
(project description)
OF
(owner's name)


TABLE OF CONTENTS

Page

1.0.  Scope of work

1.1.  These Special Conditions, together with the Agreement, General Con-
      ditions, Technical Specifications, and Drawings cover the work of (de-
      scription of work, project description, and owner's name and site location).

1.2   (Further descriptions of work scope, such as buildings, areas, systems,
      limits, and work specifically excluded from this contractor's scope.)

2.0.  Project job rules

2.1   (Insert or attach job requirements such as parking areas, brassing, safe-
*(continued on p. 122)*

**EXHIBIT 6.** *(continued)*

ty helmets with identification colors, project identification badges, off-limits areas, and access to existing facilities)

3.0. Furnished by contractor

3.1. The Contractor shall furnish all material, supplies, labor, services, supervision, tools, plant, apparatus, conveyances, equipment, temporary buildings, scaffolding, transportation, overhead, and incidental expense for accomplishing the work covered by this Contract, except the materials and services specifically named herein to be provided by the Owner.

3.2. Should the Contractor require additional facilities beyond those provided by the Owner, the Contractor shall provide his own additional facilities, such as for storage of materials, offices, change houses, or shops in the area allotted to him by the Owner.

3.3. The Contractor shall furnish the Owner with a breakdown of the Contract Price (based upon the actual cost of materials furnished and services performed) as may be required by the Owner to establish the cost of individual units of property and integration into his code of accounts. This requirement shall be satisfied prior to preparation of the initial progress estimate for payment under the terms of the Contract, unless otherwise specified by the Owner.

3.4. (The remaining articles under this section identify specific contractor-furnished items and may include some or all of the following:

Temporary power and water
Sanitary facilities
Layout of the work
As-built drawings (detailed instructions on this subject are required)
Reports, completed forms
First-aid facilities
Fencing, security

4.0. Furnished by Owner

4.1. The Owner will furnish, either directly or through others, without cost to the Contractor, the following materials, equipment, and services under the conditions described below.

(Insert owner-furnished items. A short description of location and conditions pertaining to each item should also be presented here. The following are examples of such items.)

1. Project identification badges indicating Contract number and employee number.

2. Space for Contractor's temporary building, at site, in area designated by the Engineer.

3. Water for drinking purposes, at valve at designated location at plant site. Storage or cooling of drinking water will not be provided.

4. Construction power 460 volts, 3 phase, 60 hertz at designated temporary power panels. The Contractor shall provide and maintain the required connections and cable from load center to his equipment. Power will not be furnished for heating or cooling Contractor's construction buildings.

5. Workmen's sanitary facilities, on site, to be shared with other contractors.

6. (List specific material, equipment, or services related to the work under this contract but supplied or performed by others.)

4.2. All questions pertaining to and requirements for the above items shall be directed to the owner.

4.3.* The Contractor shall receive materials furnished by the Owner on delivery conveyances at the job site, unless materials have been delivered prior to the award of this Contract, in which case Contractor shall receive materials on the ground as they occur or from Owner's storeroom on the site.

4.4.* The Contractor shall receive, unload, check, and store the materials and equipment furnished by the Owner, in cooperation with the Owner, and shall furnish the latter with a materials receiving report duly signed by the Contractor's authorized representative signifying that the materials and equipment were received in good order. Shortages, errors or damages shall be called to the attention of the Owner and acknowledgment with his signature so noted on the Owner's receiving report, or carrier's freight bill or delivery form, and a shortage form or damage report shall be completed in cooperation with the Owner. Contractor's responsibilities in accordance with the provision of Article 4.0 of these Special Conditions shall commence at the time of award of this contract. If the Contractor is unable or refuses to perform the receiving, unloading, storing, and maintenance as required herein, the Owner may perform such duties of the Contractor, either by himself or by others, and backcharge the Contractor for all costs associated therewith.

4.5.* After the Contractor has accepted materials and equipment furnished by the Owner as being in good condition and correct quantity at time of

* These articles apply when the Owner is furnishing the Contractor materials or equipment.

(continued on p. 124)

**EXHIBIT 6.** *(continued)*

delivery, he shall be responsible for their storage, maintenance and safety from loss or damage of any nature until the finished work and/or surplus materials are accounted for and accepted by the Owner.

4.6.\* The Contractor shall pay all demurrage charges on transportation equipment containing materials furnished to the Contractor by the Owner.

4.7.\* The Contractor shall pay for all damage to Owner-furnished equipment or materials accruing as a result of Contractor's unloading or transportation operations.

5.0. Schedule and sequence of work

5.1. The Contractor shall be mobilized on site on or before _____ and shall commence and proceed with the performance of the work with speed and diligence so as to ensure its completion on or before _____ .

5.2. The Contractor shall commence and complete the principal sections of the work in accordance with the following schedule:

| Description | Start work | Complete work |
| --- | --- | --- |
| 1. (list the principal sections of the work, such as | | |
| 2. structures, systems, floor elevations, areas, and | | |
| 3. equipment, with the appropriate start and end | | |
| 4. dates) | | |

(if scheduling information is required by the contractor for auxiliary or support services or systems, such "target dates" should also be included.)

5.3. The Contractor shall assist the Owner, on a periodic basis, in preparing detailed schedules of Contractor's remaining work as his performance progresses. The Owner will prepare an overall schedule for the total work incorporating the Contractor's detailed schedule. The Contractor's schedule will be revised when necessary to indicate any changes required by the Project conditions.

5.4. (If periodic progress or quantity-installed reports are required, they should be listed here.)

6.0. Drawings

---

\*These articles apply when the Owner is furnishing the Contractor materials or equipment.

6.1. It shall be the responsibility of the Contractor to notify the Owner, without delay, of any omissions, errors, or discrepancies the Contractor may discover in the Drawings. The Contractor shall in no case proceed without full resolution by the Owner of omissions, errors, or discrepancies.

6.2. A maximum of _____ copies of each required Drawing will be furnished to the Contractor without charge. Any additional copies will be charged to the Contractor at invoice cost.

6.3. Drawings furnished by the Contractor shall become Contract Drawings after acceptance by the Owner.

6.4. The following drawings are the "Drawings" that are referred to in various parts of this Contract and are considered as a part hereof:

| Drawing no. | Revision | Date | Title |
|---|---|---|---|

(List the drawings pertaining to this contract. Drawings from each engineering discipline should be grouped and identified as such, and "reference" drawings should be separately listed and identified. If a large number of drawings are used, they may be listed in a separate "List of Drawings" and inserted after the technical specifications, rather than listed here.)

7.0. Measurement and payment

7.1. Payment for Item I of the Schedule of Prices will be made on a lump-sum basis. Partial payments based on actual physical progress will be made in accordance with the performance payment schedule set forth in Article 7.8. The lump-sum payment, divided to state separately material cost and labor cost, shall include all costs of performing the work in accordance with the Contract Documents.

7.2. Payments for additions or deletions from the work included in Item I of the Schedule of Prices and occasioned by substantial changes in the Drawings will be made on a unit-price basis as defined in Item II. The quantity to be paid for, or credited, will be the net total of units added, or deleted, as determined from the Drawings or as determined from the estimated quantities if they are used in Item I of the Schedule of Prices. The unit prices shall cover all costs of performing the work in accordance with the Contract Documents.

7.3. The quantities to be paid for under Item III of the Schedule of Prices will be the total number of units of each item in place as shown on the Drawings. The unit prices shall cover all costs of performing the work in accordance with the Contract Documents.

*(continued on p. 126)*

**EXHIBIT 6.** *(continued)*

7.4. General

    1. Material costs shall include all permanent and temporary material incorporated into the work.

    2. Labor costs shall include wages and employee benefits, both employer, union and government furnished and employer wage taxes, and also include the unloading, storing, transferring, installing, maintaining, documenting, identification, and testing of material and equipment, and the training, testing, and qualifying of craft personnel.

7.5. Lump Sum—Specific Items

(Include any required clarification of the work that is covered under the lump-sum prices for each item in the Schedule of Prices; for example, the lump-sum price for subgrade ductbanks shall also include coordination of required concrete, excavation, and backfilling.)

7.6. Additions and Deletions, and Unit Prices—Specific Items

(Include any required clarification of the work that is covered under the additions and deletions and the unit price sections for each item in the Schedule of Prices; for example, the unit price for cable tray shall also include all hangers, supports, anchors, spacers, minor field fitting of tray, temporary supports and hangers, and identification.)

7.7. (Any other methods of payment should be clarified here. Examples are: per diem for manufacturer's representative, including expenses, weekly or monthly payments for storage or maintenance, credit to the owner for optional owner-furnished services, and so on.)

7.8. (A definite progress payment schedule, including percentages of total payment for each completed operation, should be given for those sections of the work that involve several operations over long periods of time.)

# CHAPTER 8

# MANAGING THE BID CYCLE

The bid cycle begins with the issuance of RFPs to qualified bidders and ends when the owner receives completed bids. It is during this period that the owner is tempted to reduce its active role in the contract formation process. Quite often, owners feel that once RFPs have been released, their control objectives are minimal—that the initiative has been transferred to the bidders, that "it's in their hands now, there is little we can do until the bids are received and we begin our evaluation." This passive role is not recommended, for there are many actions the owner may take that will improve the quality of submitted bids, ease the evaluation and award efforts, and in fact reduce the price of expected bids. In other words, the bid cycle is no different than other formation events—it should be actively controlled, or managed, by the owner to protect its best interests.

Let us begin our discussion of the owner's role during this period by considering the reasons why a bidder may price a proposal higher than expected. Common reasons for unreasonably high bid prices may fall into one or more of the following categories:

1. *Complementary Bidding.* This occurs when the bidder is not interested in performing the subject work (for any number of reasons) yet does not want to offend the owner or otherwise to be prevented from bidding on future work—to be deleted from future bidder lists. In these cases, bidders sometimes inflate their proposed price to a range that they believe will prevent award. In this way, they avoid the possibility of being asked to perform the work yet are not forced to decline the bidding opportunity.

2. *Bidder Error.* The bidder may simply make mistakes when calculating bid prices.

3. *Bidder Misunderstanding of Scope of Work and/or the RFP Documents.* Clear and complete RFPs help to reduce this possibility.

4. *Perceived Risk on the Part of the Bidder.* Any number of factors may cause the bidder to perceive real or imaginary areas of risk and to raise its bid price in order to provide for this risk. Depending upon the individual bidder's understanding of the scope of work and project conditions, not to mention its bidding philosophy, it may unnecessarily pad the price to cover risks that may not be encountered.

5. *Bidder Uncertainty.* The bidder may be quite uncertain about the proposed work, the owner's objectives and management plans, or his ability to perform as expected. In order to provide protection from uncertainty, he may increase his price beyond that which is justified.

Each of these factors leads to the receipt of bids that are priced higher than they would be under better conditions. The owner's major responsibility during the bid cycle is to eliminate these bidding factors or reduce their impact on bidder prices. This is best done through a controlled program of information dissemination—by letting the bidders know as much as possible about the bidding conditions, eventual contract terms, and anticipated conditions once construction begins. This is done throughout the bid cycle by:

1. Answering or responding to bidder questions in a clear and timely manner so that identical information is available to all bidders
2. Preparing and issuing addenda (changes to the RFP) to clarify any changes made from the time the RFP is released until bids are received
3. Extending or changing the bid due date
4. Transmitting extra copies of the RFP for use by all bidders
5. Conducting site visits and prebid meetings

Each of these steps is aimed at a complete and uniform disclosure of all information that may reduce the impact upon the prices submitted of bidder error, misunderstanding, perceived risk, or uncertainty. This includes information regarding the RFP itself, bidding conditions, the specimen contract documents, and anticipated project conditions.

As mentioned in the preceding chapter, a clear and concise RFP is essential. It is virtually impossible, however, for all aspects of a proposed contract to be reduced to written form in a manner that precludes error, misunderstanding, or uncertainty. Other communication efforts are needed.

## CONTROLLING COMMUNICATION RISKS

A common problem occurring during the bid cycle is uncontrolled, inconsistent, or undocumented communication from bidders to owner and from the owner to bidders. This results from many factors; chief among them is the owner's failure to designate one person as responsible for answering or responding to bidder questions or requests for information. The contract administrator is an ideal candidate for this duty. Although he or she may not be qualified to answer specific questions, such as those dealing with design specifications, material substitutions, construction scheduling, and the like, he should receive all incoming communications, secure an appropriate response from the owner's staff, and transmit that response to all bidders equally. All questions and other communications can then be recorded for eventual formalization through the addendum process. In this manner, all bidders are on equal footing, and information advantageous to each is available to all.

Common sense dictates that a single point of communication be established and that the bidders be told that all questions or other correspondence be funneled through that source. It is only through this single-point philosophy that control over the dynamic bidding process can be maintained. More often than not this is abused, with bidders communicating directly with designers, consultants, owner management, and others and the result being misunderstanding and confusion.

## ISSUING ADDENDA

An addendum should be prepared, reviewed, and sent to all bidders when a change or changes are made to the content or intent of the RFP. Significant clarifications should also be recorded and transmitted to bidders through addenda. In order to provide control over the addendum process, each addendum should be dated, numbered, and explicit in its meaning. An example addendum letter is shown as Exhibit 7. As shown in the sample proposal form (Exhibit 3), acknowledgment of addenda, by number and date, should be indicated with the submitted bids.

**EXHIBIT 7. Sample Addendum**

ADDENDUM

(Date)

(Bidder's name and address)

(Owner's name, project, contract description, and contract number)

ADDENDUM NO.
(addenda are numbered consecutively for each RFP, beginning with number 1)

Gentlemen:

Please refer to our letter dated (date of invitation to bid letter that transmitted the request for proposal to all bidders) requesting your bid for (contract description and contract number) for the (project name).

Our request is hereby amended as follows: (for second and following addenda, use "Our request, as amended by Addendum No. _____, is hereby further amended as follows:")

(Revisions should be made by document type, as the following examples illustrate.)

A.   The following revisions are made to the Proposal:*

    1.   Page 6 of 8, under "The Bidder requires the following temporary facilities," revise Item 1 to read "Construction Water Requirements."
    2.   Page 7 of 8, add as fourth paragraph: "A complete résumé of the proposed superintendent shall be submitted with Bidder's Proposal."

* Changes to the schedule of prices should be made by issuing replacement sheets for each page to be modified. These sheets should be identified by "Addendum No. _____" under page numbering. When several pages of the proposal are revised, it is recommended that an entirely new proposal section be issued.

B.   The following revision is made to the Special Conditions:
    1.   Page 3 of 17, Article 4.1.6, add the following:
       "8. Furnishing and installing level indicators"

The bid due date of _____ shall remain unchanged (or) is hereby changed to _____ .

Except as specifically revised by the terms of this Addendum No. 1 (or) by Addenda Nos. _____, _____, all provisions of the Request for Proposal as originally issued shall remain in full force and effect.

Kindly acknowledge receipt of this Addendum and that you have taken all provisions hereof into consideration in preparation of your bid by signing in the space provided below and returning this Addendum with your Proposal.

Very Truly Yours,

_____
(owner's signature)

Received by:

_____        _____
(name and title of bidder)                  (date)

---

Because addenda are often misused or misinterpreted, control over their use cannot be overemphasized. The following suggestions are intended to help owners avoid commonly encountered problems when using addenda to modify RFP documents:

1.  Use only one type of document to modify the RFP, and modify the RFP only through that document. Avoid multiple change instruments. When modifications or exceptions are made verbally, through letters to bidders, or otherwise, make sure they are included in an addendum issued soon thereafter.

2.  Incorporate changes made by addenda into the subsequent contract documents once a successful bidder is chosen. Do not incorporate all addenda issued during the bid cycle "by reference." This makes interpretation and administration of the eventual contract difficult. Often a plethora of addenda are issued, and it is not uncommon for one addendum to modify, enhance, or extend the changes made by previous addenda. To ask those responsible for contract administration not only to understand the actual contract documents but also to wade through all issued addenda invites misinterpretation and serious performance risk. Since the conformance of the contract represents the owner's last chance to start with a fresh set of documents, make sure this is done. Subsequent "contract changes" will make interpretation difficult enough.

3.  Attempt to make addenda brief, concise, and easily understood.

4.  Issue addenda in a timely manner. Let bidders know of changes as soon as possible so that the addenda's full impact may be understood and reflected in the bids.

5. Send addenda only to those who originally received the RFP. Do not get involved with subtier information responsibilities. Addenda should state that it is the bidder's responsibility to transmit changed requirements to other affected parties (subbidders, material vendors, and so on). Make sure that addenda are circulated internally within the owner's organization, particularly to those preparing fair price estimates.

6. Always reference the affected RFP, by section, page, paragraph, and/or line number. If a requirement is given in more than one place, make sure it is changed in each location. This is another reason to avoid repetitive or redundant requirements in the RFP.

7. Oftentimes significant questions will cause the owner to clarify but not change existing RFP material. When this is done, clarifications or enhanced explanations should be segregated from actual changes and identified as such.

8. Always refer to the bid due date, even if it is unchanged by the addendum in question. This is the first question bidders will have when receiving addenda: is the bid period extended?

9. Number addenda consecutively and always refer to previously issued addenda by number and date. This helps all parties keep track of previous changes. If a bidder failed to receive an addendum, this should be discovered as soon as possible.

10. If massive revisions to a section are needed, consider issuing replacement sheets or a new section incorporating thoses changes. To ensure that the new sections or sheets are used in the final bid, number or otherwise identify those sheets before sending them to bidders.

11. Require receipt acknowledgment. Some owners send two copies of the addenda, asking the bidders to return a signed copy upon receipt. Others ask bidders to list the addenda in their proposal. Both are recommended.

12. If addendum changes are significant, contact the bidders in advance of its issue. Let them know of changes as soon as possible, but always follow this notification with the addendum itself.

## REQUESTS FOR ADDITIONAL DOCUMENTS

From time to time owners will be requested to provide additional copies of the RFP documents, particularly the drawings. A good practice is to respond only to bidder requests—to inform subbidders to contact the prime bidder in question for additional copies. Most owners establish policies dealing with the cost of these documents, either absorbing it themselves or charging the requesting party. As with requests for additional documents, subbidder questions regarding the RFP should be directed through bidders. Once again, this is done to reduce annoying or frivolous inquiries and to maintain a lack of association with any groups other than the invited bidders.

## EXTENDING THE BID PERIOD

A decision to extend the bid period should be based on significant occurrences that justify the postponement, rather than on the convenience of a bidder. If the RFP is well prepared, a sufficient and appropriate bidding period has been scheduled, and no extenuating circumstances arise, the bid date should be held firm, despite the occasional protestation of an overworked bidder. Many contractors will request an extension as a matter of course, hoping to ease their estimating workload or to give priority to other owner bids ahead of that of the owner in question. Of course, the length of the original bid period should take into consideration the complexity of the scope of work, firmness of technical information, number and complexity of pricing alternatives, and other factors affecting the time required to prepare and submit responsive bids. Bid periods would naturally vary from contract to contract.

Some factors that may require an extension are:

1. Discovery of significant errors or omissions in the RFP.
2. Late release of the RFP to bidders or failure of any bidder to receive the RFP in sufficient time to respond.
3. Majority of bidders claim not to be able to meet the due date. It makes no sense to stand firm on the original due date if no bids will be received by then.
4. Changes in the project schedule that cause delay of the construction start-work date (bids are not needed as early as originally planned).

When these or other causes require an extension, all bidders should be notified by telephone or telex followed by a written addendum. If some bids have already been received, those bidders should be allowed to retrieve and resubmit their bids at the later date. This is another reason to refrain from opening bids as soon as they are received (rather than simultaneously on the bid due date).

## QUESTIONS OF PROPRIETY

Because the bid cycle represents the period when the owner is probably most susceptible to impropriety or the accusations of impropriety, specific precautions should be maintained. These include owner policy on contact by its employees or representatives with prospective bidders, the acceptance of gifts or gratuities, and other activity that may impugn the objective receipt and evaluation of competitive bids. Often bidders will ask to call on the owner to discuss the RFP, review drawings, or simply chat during the bid cycle. This and all other actions that affect the perception of propriety are not recommended.

The owner and its agents should maintain an arm's-length approach toward all prospective bidders and contractors. This advice pertains to the administration period as well, for the evaluation of claims or change order quotations may be impugned by familiarity beyond that which is absolutely required to conduct the business of contracting.

## PREBID CONFERENCES AND SITE VISITS

In order to ensure that all bidders are informed of the intended scope of work, project requirements, and rules of the bidding process, it is usually advisable to conduct prebid conferences and require all bidders to visit the site and acquaint themselves with conditions that may affect their performance, if selected. Site visits are handled in one of two ways:

1.  Allow bidders individually to schedule their visits with the owner's site manager and conduct independent site visits or tours.
2.  Require all bidders to visit and tour the site at a common, specific time.

Oftentimes, when the project site is under development and facilities exist, the site visit is combined with the prebid meeting. The meeting takes place at the site and is followed by an owner-conducted tour of the project, sometimes called a "job-walk." The site visit is to acquaint the bidders with such things as access, storage and laydown areas, spatial restraints (crowding), and weather. It also counters possible claims at a later date by contractors who feel they were not informed or aware of performance-affecting factors.

The prebid conference, on the other hand, is a good-faith effort by the owner to inform, clarify, or explain RFP content to all the bidders at the same time rather than going through a cumbersome process over the telephone or through the mails. All bidders are invited, along with any subbidders whom they may wish to escort. An agenda is made and distributed, and bidders are encouraged to ask any questions or request any clarification to the RFP at the meeting. So that the meeting is meaningful, it should be scheduled midway through the bid cycle, or at least after the bidders have received and had sufficient time to become familiar with the RFP documents. If conducted properly, the prebid conference is a valuable control technique for contract formation. Problems will occur, however, if it is not performed correctly. The following recommendations are intended to avoid such problems:

1.  Conduct a joint site visit. Avoid scheduling numerous visits only at bidder convenience. This assures consistency of information and is easier to manage.

2.  Prepare a strict prebid conference agenda and stick to it. Avoid extraneous, off-the-cuff remarks that may confuse bidders.

3.  Rehearse the conference. Make sure each speaker addresses only those topics assigned to him or her; if a question is beyond his expertise, defer to one who is responsible for that aspect. The construction manager should not answer design questions, and the design engineer should not answer contract administration questions.

4.  Do not allow changes to the RFP *except* through written addenda. Inform the bidders that any changes resulting from the conference that affect the RFP will be implemented through written addenda to be issued following the conference and that these will constitute the only changes.

5.  Keep control of the meeting. Do not allow simultaneous or parallel discussions. Discourage side conversations after or before the meeting.

6.  Keep minutes but do not distribute them to bidders or incorporate them into the RFP. Use the addendum process.

7.  Do not allow bidders to "dry run" their proposals. Many bidders will attempt to test your acceptance of offerings at the meeting. This should be done only during bid evaluation.

8.  Present a united front. Argue in advance as to who speaks for the owner. This will minimize having to correct one another in front of the bidders.

9.  Avoid the temptation to answer all questions on the spot. If a question cannot be answered immediately, say so. Clarify it later in an addendum.

10.  Distinguish between actual changes to the RFP (and therefore candidates for addenda) and mere questions of interpretation.

11.  Review and summarize the meeting, areas of change, expected addenda, and so on, prior to adjournment.

12.  Maintain a professional attitude at all times. Do not show favoritism toward any bidders. Avoid casual conversation and the appearance of impropriety; that is, do not go to lunch with a bidder, drive to the meeting with one, or have drinks afterward with one.

## REPRESENTATIVE CASES

1.  During a crowded and boisterous prebid conference, a question regarding sales tax is addressed to the speaker, the owner's mechanical superintendent. The superintendent, thinking of a previous project on which sales tax was paid by the owner, answers that sales tax should be excluded from the bid amount, and the discussion quickly turns to another topic before adjourning. The contract administrator, aware that on this project sales tax is to be paid by contractors, spends the remainder of the week trying to contact each bidder and correct the misstatement.

2.  As the contract administrator is entering his car in the parking lot after a prebid conference, an attendee, who identifies himself as a representative of Bidder C, stops him and states that due to the apparent confusion on the project, his company does not want to bid. The contract administrator therefore sends no further addenda or otherwise communicates with Bidder C. As it turns out, the prebid attendee, formerly of Bidder C, had just negotiated an employment agreement with Bidder B, a former competitor, at the prebid conference, and Bidder C is vitally interested in the proposed work.

# CHAPTER 9

---

# BID RECEIPT AND EVALUATION

---

When completed bids are received by the owner, two interrelated objectives should be met:

1. An impartial, objective evaluation of all bids, including such aspects as price, technical content, and commercial considerations, should be made so that the best available offering is chosen for award.
2. The security of confidential bid and bid evaluation information should be maintained at all times.

In order to achieve these objectives, the bid receipt and evaluation process should be well planned before the actual bid due date arrives. The evaluation process should be structured, with owner personnel and their representatives made fully aware of their roles. For major, complex scopes of work some owners appoint a bid evaluation committee to perform the review and make an award recommendation. But for most situations assigned personnel representing various disciplines— cost engineering, contract administration, design engineering, and so on—are appointed and perform separate evaluations—with the contract administrator acting as coordinator for this effort.

## BID EVALUATION PROCESS

A general bid evaluation process, containing separate commercial, technical, and cost reviews leading to a consensus of opinion and eventual award, is represented in Figure 15. When the RFP has been carefully prepared and the bid process managed well, the evaluation is usually a straightforward process. Elements that can complicate the effort are:

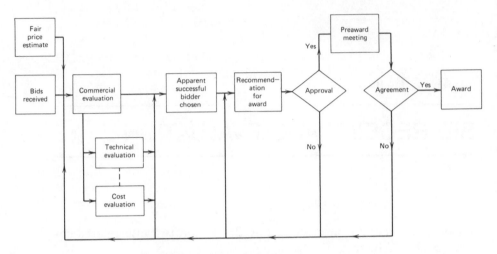

**FIGURE 15.** Bid evaluation process.

(1) ambiguous RFP and specimen documents; (2) an unmanaged bidding cycle; (3) a large number of bid alternates or complex pricing arrangements; and (4) an uncontrolled evaluation methodology. Perhaps the most common difficulty associated with this effort concerns the lack of definite responsibility assignment—where the identity and respective roles of owner or agent personnel are not clarified prior to bid receipt.

Most owners document the results of the bid evaluation on some sort of "spread sheets" that list the bidders, detail the contents of each bid, and record the data leading to an award recommendation. This is a good practice, for it is not unusual for the process to be repeated more than once for the same contract (should an agreement not be forthcoming with the apparent successful bidder). It may be necessary in these cases to return to the other bids to make a second choice. Well-documented evaluations allow this to proceed rapidly, and speed is often of the essence—valuable time has been used for the original evaluation process.

For unit-price and lump-sum bidding, the price evaluation is straightforward. Estimated units (quantities) can be used to extend unit prices to anticipated costs, and a "bottom-line" cost to the owner can be determined. Base bids should be evaluated along with optional or alternative bids. Nonprice information should also be documented on such a form. This may include the identity of proposed subcontractors, schedule durations, bonding company, or performance data of proposed equipment.

The contract administrator usually employs the aid of a technical specialist (such as the engineer responsible for the technical specifications and drawings) to review all the bids and accept or reject each depending on technical criteria without regard to price. Ideally, only those needing to consider price data should have access to it. However, some work requires that cost and technical information be studied interchangeably—such as to determine cost trade-offs between alternative systems, components, or equipment. To the extent practical, cost-sensitive information should have limited visibility.

In many cases first costs do not represent the only costs the owner incurs. Additional costs that may affect the overall decision include "evaluated costs" or "imputed costs" to cover such areas as:

1. Evaluated costs
    Equipment performance penalties or credits
    Cash timing and terms of payment
    Escalation provisions
    Optional pricing alternatives
2. Life-cycle costs
    Energy requirements
    Annual operating costs and maintenance costs
    Replacement intervals
    Expected salvage value
    Reliability (expected costs of failure)
3. Nontechnical costs of performance
    Insurance and bonding (if these are to be for the owner's account)
    Retention terms
    Schedule incentives and penalties

If any of these cost considerations are to be made, the owner should identify them, appoint the personnel responsible for determining their impact, and have objective evaluation criteria established in advance of bid receipt.

## BID SECURITY

When bidding is by invitation, it is incumbent upon the owner to see that the pricing and other confidential information contained in those bids are secure. Other bidders should at no time have access to data contained in competitors' proposals, nor should any bidder be apprised of the evaluation status prior to award of contract. Failure to secure confidential information has several detrimental effects:

1. It may serve to reduce the number of interested bidders for future work.
2. It may cause ill feelings among the owner, bidders, and eventual contractors.
3. It may result in recriminations, questions, and harassment from unsuccessful bidders who feel they should have received the award.
4. It may weaken subsequent negotiation efforts with the apparent successful bidder.
5. It may affect the impartiality of the evaluation process itself.

In order to avoid these problems, owners should issue strict bid security procedures and enforce them at all times. When others outside the owner's organization are involved in the evaluation—such as A–E firms and consultants—the difficulty of ensuring bid security increases. The following suggestions should help protect the confidentiality of bid information during the evaluation process and afterward.

1. Bids should remain sealed until all are opened on the due date. Resist the temptation to get a head start on the evaluation effort and to open bids as they are received.
2. Bids should be opened together in the presence of only selected individuals. Some owners insist on the presence of a disinterested individual as witness to the bid opening. This person also signs a log attesting to the content (and amount) of bids received. This is done as a check against improper handling by those responsible for opening bids.
3. While being evaluated, opened bids should be attended at all times.
4. When not in use, bids and evaluation material should be locked in a secure location.
5. Evaluation information should not be disclosed to others (internal as well as external) before an award recommendation is made.
6. When possible, nonprice reviews should be performed with unpriced bids (such as for technical evaluations of proposed equipment).
7. Once the evaluation is complete, bids and evaluation material should be locked in inactive files for a nominal period of time. They should not be destroyed until long after the contracted work has been completed.

## USE OF A FAIR PRICE ESTIMATE

The use of a fair price estimate, an internally prepared bid estimate, is gaining acceptance as a valuable control and checking technique used during bid evaluations. When this technique is used correctly, the contract administrator requests the cost engineer (or other qualified owner personnel) to "bid" the work using the identical RFP documents sent to actual bidders. The estimator is given the same period of time to prepare the estimate as bidders are given to bid, with no other information or insight than is provided bidders. The resulting fair price estimate is prepared on the identical pricing forms and to the identical format as actual bids, allowing it to be easily used as a benchmark against which to measure actual bids. Other benefits of a fair price estimate include:

1.  It helps to surface bidder errors or misunderstandings. A significant difference between the fair price estimate and actual bids may indicate that the RFP and/or prebid meeting did not clarify the scope of work or pricing conditions. Should all bidders have the same misunderstanding, this may not otherwise be detected.
2.  It helps the owner isolate variances or exceptions among the actual bids.
3.  It points out subjects for preaward negotiation.
4.  It identifies unbalancing of bids (front-end loading, addition and deletion unit pricing, and so on).
5.  It is a guard against potential bidder collusion or bid rigging. When actual bids are very close to one another in price, the appearance of healthy competition may be deceptive. The closeness could represent a deliberate attempt on the part of all bidders to "fix" the price.

In order for a fair price estimate to achieve these objectives, it should be prepared "blind." That is, the preparer should not see the genuine bids before preparing his or her own, and no inside information should be provided him. A common mistake, however, is to deny the preparer access to submitted bids during or after the evaluation process. These need to be seen in order to correct and update estimating assumptions and cost schedules.

## SELECTION AND NOTIFICATION

On rare occasions, a clear bid winner is determined, no exceptions or qualifications are contained in its bid, it is technically qualified, and an award can be easily made without postbid discussions or negotiations. Quite often this does not happen, and even though an apparent successful bidder is determined, specific issues must be resolved before an award is advisable. Preaward meetings are held to resolve these issues or to clarify any inconsistencies in the selected proposal.

When this happens, the bidder should be so notified and a meeting scheduled. He should not be told that he is the low bidder, or even the apparent successful bidder (most will assume this in any case). Some owners refer to these meetings as "postbid" rather than as "preaward" for this reason. If the discussions or negotiations are successful, award can then be made. Only at that time should unsuccessful bidders be notified, in letter form, that they were not chosen. The owner may or may not wish to let others know who was chosen at this time, but this will become apparent to all shortly. The specifics of the evaluation—who was second, third, fourth, by how much, and so on—need not be disclosed.

## ERRORS AND REBID

Errors in bids are common, and sophisticated owners realize that to accept a bid with admitted errors—to force the bidder to perform—would jeopardize successful performance. If errors substantially affect the price and order of bids, a rebid should be considered. The bidder in error should be offered a chance to correct its errors, and the bids should then be reevaluated, if required. This is another reason to hold off opening any bids until all have been received. It is much easier to allow bid retrieval (such as when a bidder suddenly discovers it made a price mistake) before bids have been opened than afterward.

If significant RFP errors surface after receipt of bids, or if the proposed scope of work is seriously altered, it may prove wiser to rebid the contract—send out new RFPs to all bidders and repeat the bid process—than to award a contract to the apparent successful bidder and attempt to negotiate the subsequent changes.

## REPRESENTATIVE CASES

1. The chief purchasing agent for a large electric utility had the habit of opening contract bids as soon as they arrived in order to get a headstart on bid evaluations. For a large mechanical contract, two of four bids arrived in the mail 10 days prior to the due date, and the purchasing agent had his staff begin preparing spread sheets for the evaluation. Lump-sum prices for the two early bids were:

Bidder A:  $7,298,660
Bidder B:  $8,925,000

Two days prior to the bid due date, the construction manager called for a one-week bid extension. Since no changes were made to the scope of work, Bidders A and B allowed their bids to stand. On the revised due date, the remaining two bids arrived. They were:

Bidder C: $10,100,500
Bidder D: $7,298,600

The owner awarded the contract to Bidder D, who was suspiciously only $60 under Bidder A. Bidder A complained bitterly to the utility and the press that the security of his bid was compromised. The resulting furor delayed award of the contract and start of work for four months.

2. A cost engineer for an A–E firm handling procurement for the project of a long-standing client left his desk to attend a meeting on bid security. He also left the bid evaluation sheets he was working on opened across his desk top. A competing bidder, visiting the office to distribute brochures to a buyer across the hall, happened to notice that his company's bid was low by $97,000. When he returned to his own office, the contractor's representative had his estimator call the A–E firm and report a $90,000 mistake in his firm's bid estimate.

3. The vice president of operations for a major manufacturing firm was irate when he called the project manager of its consulting engineering firm. It seems he had just received a call from a project bidder who knew the exact amount of each of the seven bids now being evaluated by the engineering firm. During the next few days, the leak went undetected, but the client stressed that it must have come from the consultant's shop, as the client, though privy to the bids, adhered to a strict bid security procedure. Though aware of the need for security, the engineering firm had no formal procedures in that regard. As a result, the client removed procurement services from the engineering firm's scope of work—at a substantial reduction in its fees for the project.

4. A major mining and minerals company would evaluate lump-sum construction bids by choosing the low bidder and having a technical evaluation made of that bid alone. On a proposal for coal handling equipment, Bidder A was awarded a furnish-and-install contract at $2 million. After five years of operation, the mining company's accountant determined the actual cost to be $5 million as follows:

Purchase price:    $2,500,000
Fuel costs:           200,000
Maintenance:          700,000

| | |
|---|---|
| Spare parts: | 400,000 |
| Down time costs: | 1,200,000 |
| Five-year cost: | $5,000,000 |

On checking further, the accountant found that another mine of the same company had bought the same type of equipment at about the same time. The second mine, however, had chosen equipment from Bidder B for $3 million. The cost experience on this equipment was:

| | |
|---|---|
| Purchase price: | $3,000,000 |
| Fuel costs: | 100,000 |
| Maintenance: | 400,000 |
| Spare parts: | 300,000 |
| Down time costs: | 400,000 |
| Five-year cost: | $4,200,000 |

5.   For an eight-mile oil pipeline project an RFP was released for furnishing and installing buried piping on a lump-sum basis. Bids were:

| | |
|---|---|
| Bidder A: | $2,100,000 |
| Bidder B: | $2,050,000 |
| Bidder C: | $2,150,000 |
| Bidder D: | $2,000,000 |
| Bidder E: | $1,950,000 |

The construction manager, satisfied that a competitive and representative set of bids had been received, scoffed at the need for a fair price estimate. Since one had been prepared, he read its amount of $1.1 million in amazement and ordered his cost engineer back to his calculator for a recheck. The result was the same. A preaward meeting was called, and Bidder E was asked to provide his estimate units of pipe, connections, cut and fill, and so on. When the cost engineer extended these numbers by the unit prices furnished in E's bid for adds and deducts at the meeting, he arrived at the price of $1 million. Bidder E, visibly flushed, called for an adjournment to telephone his office. When he returned, Bidder E explained that a "clerical mistake" must have occurred. His revised bid, he allowed, was $1.1 million. He was awarded the contract for that amount and performed remarkably well. Although never proven, an official of Bidder E later intimated to the contract administrator that all five original bidders held their own "prebid" meeting prior to submitting bids. They decided to add $800,000 to each of their bids. Without the owner's fair price estimate, the bids would have appeared competitive, and the bidders' collusion would have gone undetected— and cost the owner $850,000.

6.  In order to qualify for exclusion from clean air standards for a coal-fired electric generating plant under a grandfather clause, foundations for both 400 megawatt units needed to be solicited. Contractor A emerged as apparently successful. Contractor A was called to arrange a preaward meeting in April—eight months before the deadline for completion of the affected foundations. Since construction of the foundations would require seven months, Contractor A, knowing the extreme urgency on the part of the owner, procrastinated. It continually found reasons for postponement of the meeting, and therefore delay of contract commencement. As time passed, the criticality of schedule considerations grew, until Contractor A finally arrived at the negotiation table with time on its side. The project owner was under such pessure to begin work (a $200 million potential savings without the need for certain pollution control equipment compared with a $3 million excavation and foundation contract) that it met virtually every demand the contractor presented. Contractor A was awarded the most favorable contract in its history.

# CHAPTER 10

---

# CONTRACT AWARD

---

Once a successful bidder has been selected and both parties sign the contract documents, an award has been made. Provided no unresolved issues remain and no significant changes have occurred during or after bid evaluation, the award process is usually a simple exercise. Contract documents are prepared to reflect the specimen terms contained in the RFP as well as the contents of the proposal submitted by the apparent successful bidder.

The formal execution of the contract documents, then, signifies the end of contract formation and the beginning of the next major phase of contract management: contract administration. As simple as this exercise may seem, there are usually many tasks required to ensure that it is performed in a timely manner. And there are many pitfalls that owners should recognize and avoid during this critical period of contracting. Many of these stem from the time pressure involved—the need to mobilize and begin construction work without further delay. Owners who do not allow sufficient time for the formation period, who use poor contract and RFP documents, who do not control the bid cycle often find themselves in compromising positions during contract award. They accept imprudent terms and conditions, compromise on control and quality standards, or, worst of all, order the contractor to commence work without the protection of signed documents.

## THE ROLE OF NEGOTIATION

The term *negotiation* conjures up visions of a group of opposing parties, sitting across a table from one another and engaging in lengthy, high-stakes bargaining sessions. Many of those who have not been involved in such exercises view negotiation as a glamorous adventure, full of risks to be taken, points to be made, or sizable gains to be won or lost depending on one's "negotiation skill." In a way, this view is correct— negotiation is a highly subjective and often unpredictable gamble. As such, every effort should be made not only to enhance one's negotiating skills in anticipation of negotiating a contract agreement but also to avoid the need for negotiation entirely.

Clear, concise owner requirements (represented by properly designed and interpreted RFP and specimen contract documents), an adequate number of qualified bidders, and a disciplined bidding process are major factors in reducing the need for preaward negotiation. The more nebulous the documents, untenable the owner's position during formation, or unreasonable the owner's price expectations, the greater is the danger of relying on negotiation as a formation tool.

In general, contractors who have been in the business for any length of time are much more adept at negotiation than most owners they will encounter. They engage in contract negotiations continually— with subcontractors and suppliers as well as other owners. Most owners face negotiation of major contracts infrequently, are not sufficiently trained or skilled to achieve success, and often fail to do so. They have little to gain and much to lose. Wise owners realize these drawbacks and avoid the need for negotiation whenever possible. Besides the risk involved, negotiation seldom results in an amicable agreement, with both parties pleased with the outcome and anxious to begin their contractual obligations on good terms. Whether immediately or soon thereafter, one or both parties may feel cheated, outmaneuvered, or in a poor bargaining position for future negotiations, which are bound to take place during the course of contract performance. This is not a recommended frame of mind at the beginning of a long and challenging project.

When negotiation is unavoidable, owners should prepare themselves well. Negotiation team members should be chosen and their roles and responsibilities fully understood. They should be vested with the authority to conduct the negotiations—that is, to come to agreements, make decisions, and commit the owner when necessary. The same should be required of the contractor. Nothing is more frustrating than reaching a long-attempted agreement with the opposite party only to find that it must obtain the authority to fulfill the commitment at some higher level.

Negotiable elements of the proposed contract should be distinguished from nonnegotiable ones before the discussions begin. Owner personnel should be acutely aware of each. Their preparation should include rehearsals, the creation of a strict agenda, and discussions and/or approval of possible agreements with higher owner management, if needed in advance. Careful records should be made of the negotiation results, and these should be formally incorporated into the subsequent contract documents. Every preaward understanding should be reduced to written form and made part of the contract before the award is made.

One final suggestion regarding negotiation involves the use of bidder or contractor feedback to contract documents or formation procedures. Should an owner find itself continually negotiating on the same terms or conditions, or with the same set of bidder objections or exceptions taken to RFP documents, time may have come for a review of these terms and conditions. Some may be unreasonable, unclear, or even unnecessary. If so, they should be removed from future consideration or revised to reflect current practices in the industry. To ignore these signals—that the contract documents are, in effect, lacking—is to risk the increased time and cost required continually to clarify and negotiate from an untenable position. Quite often the terms in question do not actually offer owner protection or are not enforceable anyway. Contractors, knowing this, may accept these terms only as a compromise for the owner's acceptance of very real, meaningful, and enforceable terms favorable to the contractor.

## BID SHOPPING

Bid shopping refers to the practice of "auctioning" the contract award to the low bidder. When it is used, the owner calls in one bidder and confronts it with the lowest bid. It then asks the bidder present if "he can beat this price," with contract award hinging upon the bidder's response. If the answer is no, the owner calls in another bidder and asks the same question, until the "counteroffer" by the owner is accepted. And it is not uncommon for the owner, once a contractor has agreed to beat the stated price, to call in the other bidders for another round of this exercise, asking them if they can beat the new price. It is a ratcheting process conducted by the owner, with the price being gradually squeezed down to the lowest level acceptable to one bidder.

Most sophisticated owners feel that there is no need for bid shopping. Any minor price reduction achieved through these means is greatly offset by the serious disadvantages it presents. Among these is the common result of a contract agreement being achieved through bad faith. Contractor performance is bound to suffer when award is made in this manner. Quality is often shortchanged, and subsequent agreements for such things as change orders, schedule acceleration, and the like are difficult to achieve when the contractor feels he has "left money on the table" prior to award. Other disadvantages arise beyond the immediate contract. Owners who gain a reputation as bid shoppers will find that the ensuing bids are padded by most contractors in anticipation of the price reduction they expect through bid shopping. And many contractors may simply refuse to bid on contracts when they know bid shopping will occur. This reduces the number of bidders and therefore reduces healthy price competition.

Bid shopping is often used by contractors when obtaining quotations from suppliers or subcontractors. But owners who expect qualified contractors to bid in good faith should avoid this practice. A well-prepared RFP, adequate number of qualified bidders, and thorough evaluation process should achieve more acceptable results.

## CONFORMING THE CONTRACT

Using the original RFP documents, minus those that are for bidding purposes only—the invitation and proposal forms—the contract administrator should then prepare the actual contract documents by:

1. Incorporating all changes made through addenda during the bid cycle
2. Transferring bidder submittals and other variable data to the agreement form (prices, schedules, methods, procedures, equipment data, and so on)
3. Incorporating changes and understandings made during preaward and negotiation sessions

The resulting contract documents, together with the drawings, constitute the conformed contract. Several copies are made and sent to the contractor for signing (execution). Once returned, they are signed by the owner and distributed to the contractor and others within the owner's organization. It is recommended that the contractor sign before the owner unless both sign at the same session, for some contractors have been known to alter the documents before signing them and returning them to the owner. Detailed scrutiny of each page of the documents can be avoided by following the suggested sequence.

Often there are ancillary documents requiring the signature of either or both parties as part of the contract. These may include bond forms, project labor agreements, or waivers of lien. All such documents should be reviewed for proper completion and signature before the owner signs. As mentioned earlier, incorporation of the bidder's proposal by reference (rather than transferring necessary data to the owner's agreement form) is not recommended.

Copies of the executed documents should be filed for record in the owner's office. Conformed copies—identical to those signed, but without signatures—may also be distributed to those having a need within the owner's corporate or project organization, particularly field construction management and contract administration personnel. Since the bid drawings are now part of the contract documents as well, some owners require both parties to initial and date one or more copies of these, indicating they are the agreed drawings of record for the contract. Because a flurry of drawing revisions often follows contract award, the practice of establishing a "base set" of drawings is good. It allows contractual changes required by drawing revisions to be more easily determined when they occur.

## LETTERS OF INTENT

Although conformance of the contract documents should be a minor and rapidly accomplished chore (given proper contract formation controls), many owners are not able to produce an acceptable set of contract documents as quickly as they wish to have the contractor mobilize or begin working at the site. In these cases, resolution of any minor negotiations or the processing of paperwork is not allowed to impede contract award. An interim contract, called a letter of intent or letter contract, is issued the contractor. It references the contract, the proposal, or RFP and orders the contractor to proceed with the work pending issuance of the actual contract documents.

Even when they are used sparingly, under carefully controlled conditions, and are precisely worded, letters of intent have limited value. Otherwise, they are extremely dangerous and should be avoided, if not prohibited. They are frequently relied upon as an excuse to delay contract document execution or resolution of troublesome differences between the parties—differences that must be addressed before a contract can be signed. Indeed, some would argue that the mere existence of the possibility of a letter of intent actually encourages procrastination in resolving problems—the letter of intent relieves the immediate need for their resolution. Letters of intent should not be used as a matter of course or abused by allowing reconciliation prior to a contract to be delayed.

The major problem with letters of intent is that, for sake of brevity, they reduce the contract documents from many pages (usually several hundred) to one or two. This cannot be achieved without loss of owner (and contractor) protection. A contract thus formed exposes the owner to severe commercial and technical risks. If letters of intent are used because of a paperwork backlog, this problem should be corrected. When they are used because differences are not yet resolved, these differences should be faced. Perhaps they are incapable of resolution, and a contract between the two parties is unachievable. If so, embarking on an unachievable contract with scant protection is the greatest of errors.

For those owners who insist on using letters of intent under these circumstances, the letters should be carefully drafted. They should be numbered, be dated, incorporate the accepted terms and conditions of the RFP by reference, and be countersigned by the contractor. If any bonds are required, they should also be received. Insurance certificates should be submitted by the contractor before commencing work. Scope and price limitations are also strongly recommended and should be specified in the letter. All letters of intent should state that they are temporary in nature and be effective for only a certain period of time, such as including an expiration date 30 or 60 days later. If the letter of intent is about to expire and a contract has still not been drawn and signed, an extension of the letter can be issued—in effect for another stated period of time. This achieves two important objectives: it limits the owner's cost exposure, and it forces owner management to review periodically the conditions retarding proper formation. The pending expiration serves as a "tickler" to those responsible for resolving the issues in question and executing more adequate contractual documents.

Despite all the risks associated with letters of intent, their use persists. It is not unheard of for an entire contract scope of work, spanning one or more years to complete, to be performed under a letter of intent—that is, an executed contract is never achieved. The goal of those responsible for contract formation should be to schedule and control their activities so as to eliminate the need for and the exposure caused by letters of intent. These instruments probably represent the single most dangerous practice associated with the entire subject of contract formation.

# PART 3

---

# CONTRACT
# ADMINISTRATION

---

As opposed to the activities of contract formation, where a fairly linear sequence of events takes place for each contract to be awarded, contract administration represents a collection of processes, responses to events, and controls that are not as easily arrayed or performed. Some requirements of contract administration are common for all contracts, such as progress payments, record keeping, and contract closeout. However, there may or may not be a need for other business transactions and controls. Claims, change orders, backcharges, and brief-form contracts represent some topics in this category.

This part will present the broad topic of contract administration by discussing both sets of circumstances: those that occur for all contracts, and those that typically (some would say unfortunately) occur during the contract performance period of many contracts. Major topics include:

*Mobilization and Commencement.* Initial meetings and communications between the contracted parties will be described, along with a system for maintaining critical contract documentation, correspondence, and submittals.

*Progress Billings and Payments.* Periodic progress payments are a way of life in the construction industry. We will discuss ways to ensure that they are objectively determined, in compliance with the intent of contract documents, retain owner leverage over work to be performed, and are processed in an efficient and timely manner.

*Notifications and Change Orders.*   Nothing is more constant than change in construction work. This chapter will describe typical causes of change and ways in which an effective change control program may be implemented.

*Backcharges.*   This chapter describes the typical conditions provoking the need for backcharges, the procedure recommended for cost recovery, and ways in which backcharges may reduce the total project cost.

*Claims.*   Claims will be discussed from the owner's perspective, whether it is (1) defending against a construction claim brought by a contractor; or (2) pursuing a claim against another party. Claim causes will be analyzed, and typical components of contractor claims will be addressed.

*Brief-Form Contracting.*   Most sophisticated owners allow a certain degree of flexibility needed to respond to unforeseen events through the use of brief-form contracts with certain scope or dollar limitations. This chapter will describe the use, abuse, and processes by which brief-form contracts are employed.

*Contract Closeout.*   This chapter addresses the wrapping up of unfinished business upon contract termination. Final payment, delivery of remaining products and information, and postperformance controls will be presented.

Contract administration begins with the signing of contractual documents, continues throughout the performance period, and ends with the formal termination of the contractual relationship. Although the success of contract administration activities depends in part on the relationships, agreements, and controls created during the formation period, much can be done during the administration phase to ensure project success and reduce commercial risks.

# CHAPTER 11

# MOBILIZATION AND COMMENCEMENT

Once a construction contract has been awarded, two physical mobilizations begin. The contractor begins to gather the people, material, and equipment it will need to commence work at the site, and the owner should in turn mobilize the personnel, systems, and controls required to manage the contractor's performance. For the contractor, this mobilization requires planning, assignment of personnel, and acquisition of hardware such as construction equipment and materials, and the acquisition of funds, bonds, insurance, licenses and so forth. It is an exercise in logistics and, depending upon the scope and location of his work, could represent tremendous effort. The owner's mobilization is also a logistical challenge. By the time the first contractor begins site work (ideally before it arrives at the site), the owner should be prepared to perform the twin functions necessary to protect its own contractual interest: technical monitoring and contract administration. This chapter focuses on the initial stages of the latter.

## MAJOR ACTIVITIES

Once the contract administrator has been selected and given an executed copy of the contract documents, he or she should begin to acquaint himself or herself with their contents. Since his job is to ensure contractual compliance of a commercial nature, he needs to become intimately familiar with the requirements of both parties, owner as well as contractor. Because it is usually inconvenient to wade through actual contract documents in search of specific requirements—made even more difficult when they are located in several documents or are buried under nonrelated material—most experienced contract administrators prepare contract abstracts. These abstracts reduce the voluminous contract documents to their essential elements—the commercial requirements of both parties. When such compliance aids are used, however, the contract administrator should make absolutely sure they reflect all requirements accurately. In addition, as the contract documents are amended from time to time, such as through change orders, so should be the contract abstracts.

Once the contract administrator has become familiar with the contract(s) he is to monitor, and before the contractor(s) begins substantive work, three initial tasks must be completed:

1. An initial set of contractor submittals must be received, reviewed, and filed.
2. An initial meeting between the contract administrator and the contractor's job-site representative is held.
3. The system and facilities for maintaining contract records is established.

## CONTROLLING CONTRACTOR SUBMITTALS

Most contracts call for the contractor to submit certain items or documents for the owner's review and approval prior to commencing specified portions of the work. These are generally submitted during the initial phases of contract performance. For some contracts, particularly those involving complex, large scopes of work, these may amount to a mountain of paperwork. In order to prepare for receipt and review of contractor submittals, the contract administrator should review his contract documents or abstracts and create a list of documents he should receive. Typical submittals required by many contracts include:

1. Evidence of insurance coverage (insurance certificates)
2. Fabrication–installation procedures (rigging methods, weld procedures, concrete placement schedules, soil compaction methods, and so on)
3. Quality assurance or control program documentation
4. Shop drawings
5. Material certificates (such as for stainless steel)
6. Fabrication, shipping, and construction schedules
7. Equipment tagging procedures

8. Equipment operating and maintenance manuals and instructions
9. Signed bond forms
10. Sample products (wall coverings, floor coverings, paint chips, siding samples, and so on)

Although the contract administrator should not be qualified or responsible for reviewing the adequacy of all contractor submittals, as technical submittals are usually received and controlled by other project personnel, he or she should be responsible for monitoring the status of each and corresponding with the contractor in this regard.

In order for the contract administrator to ensure that all contractor submittals are properly received as required by the contract, a submittal checklist (Exhibit 8) is maintained for each contract. The checklist should show the contract number, contractor's name, submittals required, date required, and date received and indicate when resubmittal is necessary and resubmittals have been received.

For contracts involving detailed design on the part of the contractor or the contractor's material suppliers, hundreds or even thousands of shop drawings may be required. Quite often, these need weeks or even months to be reviewed, are often rejected and returned to the contractor,

---

**EXHIBIT 8.  Contractor Submittal Checklist**

### CONTRACTOR SUBMITTAL CHECKLIST

Project _____          Contract Number _____

Owner _____          Contractor _____

| Submittal | Date Required | Date Received | Resubmittal Required | Resubmittal Received |
|---|---|---|---|---|
| | | | | |
| | | | | |
| | | | | |
| | | | | |
| | | | | |
| | | | | |
| | | | | |
| | | | | |

and are resubmitted time and time again until finally approved. Without contractual compliance and tracking aids, control of the status of such submittals becomes impossible. In addition to contract abstracts and log sheets or checklists, use of suspense files (sometimes called "tickler" files) is recommended. These are especially important when controlling the commercial compliance of the owner. Many contractor claims result from the owner allowing submittals to languish unreviewed or undue delay in approving others. The contract administrator needs to stay on top of the receipt and acceptance status in order to prevent or defend against these claims—much less to ensure that contractor performance is not delayed through the owner's inability to turn these submittals around within a reasonable time period.

## SPECIAL INSURANCE REQUIREMENTS

Because of the extreme exposure caused by a contractor or subcontractor operating without insurance coverage, or with inadequate coverage, specific cautions regarding verification of this coverage are in order. The contract administrator must make certain that certificates of insurance are submitted by each contractor performing work at the site prior to commencement of such work. When received, these certificates should be checked against the contractual insurance requirements. It is surprising how many times these certificates are in error, are unsigned, or do not comply with coverages specified in the contract documents. New insurance certificates are often required, and their submittal should be expedited.

Signatures on insurance certificates are often made by insurance salespersons, agents, or others who may not be able to bind the actual insurance carriers or even to attest to the existence of coverage itself. These signatures should always be verified as to authority and accuracy. Anyone can sign a certificate—the contract administrator should ensure that signatures actually represent the coverage stated. In order to control insurance coverage throughout the project's duration, the contract administrator should use an insurance certificate log (Exhibit 9). Both contractors' and their subcontractors' coverage should be monitored, and care should be taken to make sure that coverage does not expire midway through the performance period. Again, suspense files are helpful in this regard.

One final recommendation dealing with insurance is offered. Because the practices of the insurance industry vary frequently, and policy limits sometimes fail to cover increased costs due to price inflation,

**EXHIBIT 9.   Insurance Certificate Log**

INSURANCE CERTIFICATE LOG

Project _____

Owner _____

| Contract No. | Contractor | Subcontractors | Date Certificate Received | Expiration Date |
|---|---|---|---|---|
|  |  |  |  |  |
|  |  |  |  |  |
|  |  |  |  |  |
|  |  |  |  |  |
|  |  |  |  |  |
|  |  |  |  |  |
|  |  |  |  |  |
|  |  |  |  |  |
|  |  |  |  |  |
|  |  |  |  |  |
|  |  |  |  |  |

insurance provisions contained in standard or reference contract documents should be reviewed periodically. It is not uncommon for these to become quickly out of date or to bear little resemblance to present insurance practices.

## MEETING THE CONTRACTOR

It is good practice for the contract administrator to schedule an initial meeting with the contractor's representative—his or her counterpart—before any significant work is accomplished or any contract administration transactions take place. The purpose of such a meeting is to establish the groundwork for good-faith relations throughout the performance period. In addition, much potential confusion or delays may be avoided when processing such items as progress payments and change orders if the procedures for handling such are reviewed by both parties before they occur. Additional procedures that the contract administrator might consider reviewing are those to be used for backcharges, verification of cost-reimbursable charges, invoice format, notifications, and contract closeout.

Whether the contractor's representative is called the project manager (contractor's), job superintendent, job manager, or any other title, he or she should be the one responsible for commercial transactions that will take place. He should not be merely the contractor's marketing representative or a home office salesperson. He should be on the site, close to the actual construction performance. Just as the contractor should expect not to have to deal with several owner representatives for common transactions, so should the owner have one individual with whom to deal when commercial matters are involved.

Another good idea is for the contract administrator to prepare sample forms, or actual ones, for commonly anticipated transactions—price quotations, change order requests, progress estimates, and so on—and to review these with the contractor's representative. Extra copies should be given for his future use. This will ensure that owner-designed forms are used wherever possible—greatly streamlining future transactions.

If other owner personnel will be working with the contractor, they should also attend such a meeting. This may include the construction manager, resident engineer, and construction specialist who will be monitoring the contractor's technical performance. The agenda for such a meeting may include some or all of the following topics:

1. *Contractual Authority.* Who may speak for the owner and the contractor, by name, for certain commercial and technical issues.

2. *Contractual Communications and Correspondence.* Numbering systems for letters and other documents, addresses, titles, and so on.

3. *Submittals.* The contract administrator may wish to give out copies of his abstracts or checklists, providing the contractor understands they do not take the place of or modify the original contract requirements.

4. *Changed Work Processes.* Notifications, change order quotations, change orders, and so on (see Chapter 13).

5. *Progress Payments.* How progress will be measured and paid for. "Master" progress payment forms should be given to the contractor.

6. *Contract Closeout.* And related terms or practices such as retention, credits for owner-furnished items, and release of liens.

7. *Backcharging Procedure.* How notification will be made, contractor and owner obligations for maintaining cost records, and so on.

## ESTABLISHING CONTRACT RECORDS

The importance of accurate, easily retrievable contract records cannot be overemphasized. Although contract filing systems may seem a clerical responsibility, they are too important to leave solely to clerical personnel. Even for the simplest of projects and contracts, the amount of contract-related paperwork needed to execute performance and control contractual compliance is sometimes staggering. Without a tightly structured and disciplined approach to the management of these records, contract administration, and indeed project management, is agonizing for all parties involved.

Figure 16 presents a recommended contract filing system for typical documents, reports, and records used throughout the formation and administration periods. Although the contract administrator should concentrate on control of performance-related material once the contractor begins work, he or she should also have information and records concerning preaward events as well. The following suggestions are given to those who may be reviewing or establishing contract record systems:

1. Avoid redundant or duplicate files. Changing or updating is often required, and this can be uncontrollable if changed documents are located in more than one place.

2. Appoint one person responsible for file maintenance and security. The contract administrator is a likely candidate.

3. Control access to and removal of material. Use sign-in–sign-out cards or other methods. Records often disappear.

4. Maintain original contracts in more than one location to guard against loss by fire or other means.

5. Maintain one complete set of contract drawings as they existed at time of award. Resist the temptation to use these for purposes other than determining a contractual or pricing basis. Do not mark them to show changes, as-built conditions, and so on; use other sets for these purposes.

6. Keep contractors away from owner's confidential material. This includes fair price estimates, other bids, and claim-defense documents. In the cramped quarters of many construction offices or trailers, this is difficult to do.

7. Avoid the temptation to "give everyone a copy of everything." Copying and distribution chores can easily become a nightmare. Periodically review contract-related distribution lists to purge names or positions that have no need to receive copies.

**FIGURE 16.** Typical contract records.

162

## THE IMPORTANCE OF GOOD CONTRACT RECORDS

Many times contract record systems are set up in haste, with little planning or regard for their further use. Additionally, the commencement period is usually one in which both parties are in good spirits and an amicable atmosphere prevails. This often lulls owner and contractor personnel alike into neglecting the defensive use of records (or, for that matter, the offensive use of paperwork) for a later, less friendly stage of the relationship. Contract records should be established with a view toward not only expected events but also potential needs. Some of the major uses of contract records are:

1. *To Ensure Technical and Commercial Compliance by Both Parties.* Records point out the need for compliance, help control compliance, and demonstrate that compliance, or lack thereof, has been accomplished.

2. *For Contract Auditing Purposes.* The expense, time required, and effectiveness of cost and operational auditing are greatly dependent upon the quality and accessibility of contract records.

3. *For Control of Correspondence.* Contract-related letters, memoranda, telephone conversations, and other correspondence are usually voluminous and may have far-reaching impacts.

4. *Change Control.* In order to determine the impact of proposed or enacted changes, some means of determining the original or current position must exist.

5. *To Make Progress Payments.* This is particularly true for cost-reimbursable work, where evidence of contract charges are required. It is also important for other pricing methods.

6. *For Quality Control.* Material certificates, erection procedures, test results, and so on, are some of the records needed for this purpose.

7. *To Make Final Payment.* Evidence of completion and acceptability is required as a prerequisite for contract closeout and final payment. Sometimes it represents events or work that occurred months or even years ago.

8. *To Operate the Completed Facility.* In their zeal to begin and complete the construction process, owners sometimes fail to remember the purpose of the completed facility and operating or maintenance needs. Typical records required for this purpose include operating and maintenance manuals, equipment wiring diagrams, as-built drawings, warranties, and training material.

9. *As a Claims Defense.*   An old maxim holds that he or she who is best documented often prevails in the claims arena.

10. *To Determine What Transpired in the Past.*   Because most projects of this nature require a long time to construct and many people associated with the beginning of a project are often long gone by the time it is completed, these records represent the collective memory of what transpired. The records should be prepared and maintained with this goal in mind.

# CHAPTER 12

# PROGRESS BILLINGS AND PAYMENTS

Periodic progress payments are a way of life in the construction industry. Unless the performance duration of a particular contract is shorter than 30 days, intermediate determinations of progress will be necessary. Prompt and accurate submittal of contractor invoices for partial payments and timely payments by the owner are critical to the financial status of most contractors. The cost of financing labor, material, and equipment during the construction period is generally too severe to expect a contractor to bear it alone. In addition, payment to its material suppliers and subcontractors is predicated on progress payments to the contractor by the owner.

Contractor concerns in this area are fairly easily seen—they wish payment of the most money in the shortest possible time. There are also significant concerns and risks for owners. Because payment represents the single most powerful compliance lever, owners should ensure that progress payments are determined objectively and are designed to promote timely performance of contractual obligations on the part of the contractor. Early payment and overpayment for contract work as it proceeds are two risks that should be avoided by the owner. Cash-flow considerations and the high cost of financing dictate that owners should pay only for what has actually been accomplished and not before it has been accomplished. On the other hand, owners who unreasonably delay or withhold earned payments run the risk of contractor dissatisfaction, which can be manifested in reduction of quality, decrease in working forces and equipment, or eventual bankruptcy. All have a negative impact on successful project completion.

## ALTERNATIVE BASES FOR PARTIAL PAYMENTS

There are three general bases upon which partial payments may be made: (1) cost; (2) time; and (3) actual performance, or progress. Of the three, payments based on progress are by far the most preferred from the owner's point of view. Examples of each are given in the following paragraphs.

1. *Payments Based on Cost.* The owner contracts for fabrication, delivery, and installation of a large turbine generator. The payment method is cost plus fixed fee. Estimated costs and fee total $20 million. Each month the contractor submits invoices representing his actual costs for material, labor, equipment, overhead, and so on, incurred during the previous month. After verification for accuracy and allowability, the owner reimburses the contractor for these costs. The fee portion of the contract price is paid in monthly installments or as a lump sum at contract closeout.

2. *Payments Based on Time.* The same turbine generator is bought for a lump-sum price (for furnishing and erection combined) of $20 million. The contractor promises to complete his scope of work within 20 months from commencement. A payment schedule is created calling for 20 equal monthly payments of $1 million each.

3. *Payments Based on Progress.* The same turbine generator is bought for a lump-sum price of $20 million, but a progress payment schedule is created with several payment milestones, each weighted in proportion to its respective value. The total value of all intermediate milestones is equal to the lump-sum price. Payment for each milestone is made upon achievement of the actual event in question. Examples of the milestones included are receipt of raw materials, completion of turbine rotors, completion of turbine stators, fabrication of the generator, release of the turbine generator for shipment to the construction site, receipt at the site, and intermediate installation milestones, ending with testing and final acceptance.

Of the three methods, partial payments based merely on the passage of time are probably the least desirable. They offer absolutely no positive incentive for contractor performance. Sometimes time-based payments are proposed where the monthly payment amounts are representative of the "planned" performance during the contract duration; that is, monthly payments are not equal but fluctuate depending upon the amount of work or cost expected as the work progresses. Even in these cases, a fixed payment schedule provides no performance incentives. When a payment schedule is created to approximate planned progress, it will probably soon be out of synchronization with actual progress should material shortages, labor disruptions, or production changes occur.

Cost-based payments bear a closer relationship to actual progress if one considers that the expenditure of costs is proportional to the amount of work being performed. But even this relationship is indirect and often out of proportion. For large material and equipment items, for example, the cost of the item itself far outweighs the labor cost of having it installed. The incentive is for the contractor to ship the item to the project site quickly, yet little cost incentive remains to have it installed. From the owner's point of view, the majority of progress is achieved upon installation. In addition, contractors interested in increasing their early payments may order or buy as much material and equipment as possible, regardless of when its installation is needed. This loads the site with material long before it is needed, causing storage and protection costs to increase unnecessarily. It is difficult to avoid cost-based progress payments when working under cost-reimbursable contract pricing.

Partial payments are helpful from a control perspective only when they make it beneficial for the contractor to perform segments of the work on time and in a quality fashion, for it should be only then that payment is made for the work in question. One can think of performance-based progress payments as cost incentives to the contractor within the overall framework of risks and incentives determined by the particular contract pricing scheme employed. Even if no absolute cost incentives are called for in the contract pricing method—as with a firm-fixed-price agreement with no performance, cost, or schedule incentives—relative cost incentives may be built into the payment process. The amount of the contractor's payments will not vary (no more or less than the lump sum is paid), but the *timing* of those payments may vary considerably. Because the time value of money is deeply felt by all contractors, timing of progress payments becomes a valuable control tool in the hands of the owner.

Performance-based payment terms are also valuable because they may place significant importance on certain elements of work that, although critical to the owner, have little cost or value in and of themselves. Testing is a good example of one such item. Though the contractor's budget for completion of testing may be small, the owner may weight this activity heavily in its progress payment scheme, prompting the contractor to complete testing as soon as possible in order to begin using the finished project or component.

In most cases, the desires of owners and contractors are in exact opposition when weighting the milestones or elements of work. The owner places heavy dollar emphasis on later stages of accomplishment to ensure their completion, whereas the contractor prefers the contract pricing to be slanted toward the front end of the job, increasing his cash flow and reducing his financing costs. This desire is the genesis of "front-end loading," of which contractors are often accused when pricing work elements in their bids.

Because they represent the most powerful compliance control, progress-based payments (or progress payments) will be emphasized in this chapter. Since payments under this scheme rely on thorough planning of payment elements and values, the incorporation of specific terms for measurement and payment in the contract documents, and careful administration controls, they should be considered throughout the contract planning and formation periods rather than only during contract administration.

## DIFFERENT REASONS FOR MEASURING PROGRESS

There are other reasons, however, for owners to be concerned with contract or project progress. These may include the owner's need to be informed as to the planned completion date, of schedule performance, anticipated project costs, or estimates of the total project upon its completion (forecasts). These are equally important reasons for measuring progress, but in many cases they are based upon criteria different from those embedded in contract payment schemes.

Sophisticated owners of large projects in the construction stage often employ project planning and control methods and systems (often computer based) in order to assess current progress and forecast future progress in terms of time and costs. These support management information needs through reporting of cost and schedule status and performance to date and forecasts of expected performance, productivity, or trend analysis. To the extent that progress measurement for these purposes uses the same milestones, criteria, or weighted values as do contract payment schemes, these two objectives may be met through use of the same techniques and data. However, it should be remembered that most owners have two distinct reasons for measuring progress: (1) to determine where the project has been, where it is, and where it is going in terms of cost and schedule; and (2) to facilitate required payments to performing contractors in a manner that promotes timely performance of contractual obligations. To force a progress-measurement technique or system designed to accomplish one of these objectives on the needs of the other is not recommended.

## THE CHALLENGE OF OBJECTIVITY

One of the most time-consuming, confusing, and often inaccurate ways to measure progress for a construction contract is to make a subjective determination based on the experience of the measurer or the visible achievement at the job site. Many contract documents state that a "mutual determination of progress and commensurate payment will be made at the end of each month," without specifying the criteria for that determination. This usually results in arguments between the contractor and owner representatives as to how much has been accomplished and how much it is worth. It is a dangerous practice for both parties—much like gambling—for each could win or lose.

In order to avoid the need for subjective determinations, objectivity must be planned during the contract formation stage and implemented during the performance period. Owners can plan objectivity through the use of clear and concise measurement terms in the RFP and contract documents. Contractual scope of work should be broken down into smaller, more discretely priced segments, each readily determined and separately priced. When this "line item" list is inserted into the RFP, contractors can plan their cost performance and bid competitively for each payment item. Once a successful bidder is chosen, the payment items are transferred to the contract documents and form the basis for objective measurement of and payment for progress during performance.

An alternative is to subdivide the work for pricing purposes after the contract has been awarded. This calls for the mutual preparation and acceptance of a "schedule of values" that totals the contract price. Although a schedule of values provides for objective progress measurement and payment, these are seldom achieved without the injection of subjectivity by both parties. And because the contract has already been awarded, the contractor is usually in the strongest bargaining position. The benefit of competitively bid payment terms is lost with this approach, and it is not recommended unless the preaward criteria have not, for some reason, been achieved.

Two types of terms regarding progress payments are needed in contract documents. The payment items and their prices should be established; this is generally done in the proposal section of the RFP and subsequently in the agreement section of the contract documents. In addition, specific measurement and payment "terms" are necessary. These describe each payment item and the methods by which it will be measured and paid. These clauses are generally contained in the special conditions, for they vary with each contract according to the scope of work, payment method, and owner control objectives—not to mention the terms that the contractor may insist upon.

Once discrete payment items are identified and weighted as to price and their measurement and payment are defined in the contract documents, objectivity must be ensured throughout the performance period. This is done through complete understanding of pricing and payment conditions by both parties, especially by those persons who will be involved in requesting progress payments (contractor) and approving or making them (owner). In addition, control and understanding of the impact of contract changes must be accomplished as the work progresses. Specific provisions must be made for pricing and payment of such items as change orders and backcharges, which may be encountered.

Of all suggestions made regarding objective progress payment activity, none is more significant than the following: Progress measurement and payment terms and practices should reflect the way in which the actual contract work will be performed. To do otherwise invites difficulty if not disaster throughout the entire payment process. Those responsible for choosing separately priced elements of work, defining milestones for payment, and weighting the value of each should be familiar with the expected method of fabrication, shipment, installation, or erection involved.

Suppose an owner wishes to promote timely completion of a three-mile-long pipeline to be buried underground. It may decide that the pipe should be paid for in sections, say per foot of pipe installed. Payment terms may require acceptance of each foot of pipe before the equivalent installation price is paid. But these two criteria are conflicting, since the contractor cannot gain acceptance of the first foot of installed pipe until the three-mile system is completed and tested for leaks. The pipe will not be installed, connected, backfilled, and tested in one-foot segments! This is an example of progress payment criteria developed in a vacuum, without knowledge of planned construction techniques. It often occurs when personnel responsible for contract formation (the development of payment terms) are not experienced in the other major phase of contract management—administration.

Another example of this problem might involve the installation of concrete in a massive foundation. The contract could call for payment on a "cubic-yard-installed" basis. However, the contractor must expend a great deal of effort and cost before the first cubic yard of concrete is poured. It will have to excavate the entire foundation, construct formwork, and install reinforcing steel and embedments. Each of these steps involves physical progress yet according to the envisioned payment scheme represents no progress. Contractors encountering such preposterous payment schemes should immediately object to their use.

For some contractor activity, particularly that involving services, it is difficult to determine products or to compare output to resources or costs expended. For these particular elements of work, some subjective measurements may have to be made. An example might be guard services provided under contract. There is no way to measure the performance of this service in a quantifiable manner, for the amount of labor expended to guard the facility is not directly proportional to any "products"—so many guards will work for so many hours regardless of the number of thefts or security violations prevented. For these items, cost-based or time-based payments are unavoidable.

## SPECIFIC MEASUREMENT TECHNIQUES

Let us examine the most commonly used measurement techniques. Although they have different names, all are similar. They each require the preperformance identification of work steps or products during the formation stage and the assignment of a dollar (or percentage of the total price) value to each. Each of these weighted "accomplishments" should be defined so that field personnel can easily determine whether they have been successfully achieved.

### INTERMEDIATE MILESTONES

This is probably the most often used measurement technique, particularly for fixed-price contracts. Each specific portion of the work is broken down into its components so that the completion of a component can be readily identified. Each component ("work item" or "line item") is given a value, and that amount is paid when the item has been accomplished.

### Example

A $30,000 contract price for the foundation of a small building may be broken out in the following manner:

| | |
|---|---|
| Foundation excavation: | $5,000 |
| Concrete footers: | 2,000 |
| Base mat: | 6,000 |
| South wall: | 2,500 |
| East wall: | 2,500 |
| North wall: | 2,500 |
| West wall: | 2,500 |
| Interior supports: | 2,000 |
| Ground-level concrete deck: | 3,000 |

| | |
|---|---|
| Cleaning and dressing and | |
| Final acceptance by owner: | 2,000 |
| Total: | 30,000 |

Provided the sequence of construction follows the preceding payment sequence, this may be a proper progress payment approach. Should the sequence of construction call for the simultaneous pouring of all four walls and the base mat, however, another more representative breakdown would give better results. Once the work has started, owner personnel would check the progress at the end of each payment period, determine which of the above steps have been accomplished, and make payment accordingly.

Should any of the preceding steps (or products) be partially complete, the payment terms may not allow partial payment. For example, if at the end of the second month the north wall is 90% complete, no payment of the $2,500 allocated for that wall is made. An alternative approach is to allow partial payments for partially completed milestones. The determination of degree of completion would then revert to a subjective one—for example, is the north wall 90% complete, or is it 80%?

When using the intermediate milestone approach, owners and contractors should ensure that the milestones are easily identifiable and that their accomplishment can be achieved in a relatively short period of time. A $50 million contract containing only four intermediate payment milestones (spaced six months apart) would not be practical. Milestones should be selected that allow the payment for progress to approximate closely, or follow, the actual accomplishment of progress. It should be remembered that the term *milestone* when used in this context refers to an identifiable payment item rather than to a significant event in the overall project schedule ("schedule milestone").

## EQUIVALENT UNITS

This technique is also common. It relies on the selection of performance or product units that are indistinguishable from one another. Each is given the same value, and each should therefore require the same resources for achievement. Examples of equivalent units may be feet of piping installed, cubic yards of earth removed, or tons of structural steel erected. It is easy to see why this technique is common for unit-price contracts. At the end of each payment period, a determination of the number of units performed is made. The payment price is calculated by multiplying the "unit payment amount" by the quantity of units.

Although lending itself to unit-priced agreements, this technique may also be applied to lump-sum pricing schemes. Lump-sum line items in the contract may be divided into equivalent units, with partial payments of the total lump-sum amount made in proportion to the number of units performed.

## Example

A lump-sum contract for $1 million is awarded for construction of an office building. Of this amount, $50,000 is allocated to the foundation excavation, and an estimated 20,000 cubic yards of earth must be removed. Payment is then made on the basis of $2.50 per cubic yard of soil excavated. At the end of the first payment period, measurements show that 8,000 cubic yards have been excavated. Therefore, $20,000 is paid for the partially completed excavation.

Significant problems develop when this technique is applied to units of work that are not exactly equivalent. For example, the equivalently priced unit may be cubic yards of concrete poured. If the scope of work calls for a base mat foundation as well as several elevated concrete decks, the cost and time required to place each cubic yard of concrete in the base mat will be far less than that required to pump each cubic yard of concrete to decking on higher floors. Even though each cubic yard of concrete is indistinguishable from the others—regardless of where it is placed—economy of scale and ease or difficulty of placement will make units installed in one location different from those installed in the other. For situations such as these, separate unit payment amounts should be considered. The same situation arises when dealing with piping. It goes without saying that 1-inch-diameter copper tubing should be priced and paid for differently than 10-inch-diameter carbon steel pipe. Again, persons responsible for designing the payment scheme should be knowledgeable in the products, methods to be used, and general construction sequence in order for payment provisions to match actual construction conditions.

A final caution applies to equivalent unit techniques when used for payment purposes. Should the achievement of the first products (payment units) be accomplished only after a great deal of preparatory effort and expense, a milestone technique should be considered. For concrete placement, a milestone scheme calling for formwork, reinforcement, embedments (anchor bolts, pipe sleeves, plates, and so on), and actual concrete pours may prove more representative of actual work than one calling for payment of only the cubic yards of concrete placed.

## THE 50–50 TECHNIQUE

This technique allocates half of the payment amount to commencement of the item in question and the remaining amount upon its completion. No intermediate measurements are made; that is, when the item is 90% complete, only the 50% allocated to its commencement is paid. This technique has limited application to major construction efforts. It is used for small items and items for which intermediate progress determinations are difficult or impossible to make. For some payment items, the technique will overestimate actual progress—as soon as the work on the item is started, it is not necessarily half finished. For other items, the technique will underestimate actual progress, as with the case where 90% is finished but only half credit is given. When many small items are measured by this method, chances are that the overestimated ones will balance the underestimated ones to some degree and that a grossly accurate representation of aggregate progress will be made.

When the 50–50 technique is used, care should be taken to ensure that the contractor is not encouraged to start work on all items—in order to capture the initial 50% of their value—with little incentive to complete the first item. In addition, the items in question should be ones that can be accomplished in short periods of time. Otherwise, this technique may give unacceptable results. In general, then, the 50–50 technique is a form of milestone technique, with the start and finish of each item representing milestones of equal weight (50% of the payment amount).

## THE 0–100 TECHNIQUE

This is a further adaptation of the milestone technique in which there is only one milestone for each payment item—completion. No payment is made until the item in question has been completed, and full payment is made at that time. Again, as with the 50–50 technique, this method should have limited application and be restricted to minor elements of work that are completed soon after they are begun.

Intermediate or final testing may be examples of the proper use of the 0–100 technique. If the line item in a lump-sum contract for equipment installation allows $500 for testing, this may be paid only upon successful completion of the test involved. It would be difficult to determine any intermediate progress on the test, and the owner is generally concerned with a successful test rather than with an unsuccessful one.

The installation of minor devices or instruments may also be priced in this manner. For example, if 100 pressure gauges are to be installed on a complicated piping system, and if the installation of each requires four hours, there is no need to determine partial progress for each gauge. The owner would simply count the number of gauges installed at the end of the payment period and pay for the value assigned to each.

The 0–100 technique alternately may be called the "all or nothing at all" method. Another example of its use is the payment for material or equipment as it arrives at the job site. Of the material cost of each item, 100% is paid as the item is received, and it is meaningless to estimate the partial receipt of a single item—for example, if it is in transit, is it 20% there, 45% there, or 90% there? The owner simply determines if the material in question has arrived or not. If so, the associated price is paid; if not, the material price is not paid.


## SUBJECTIVE TECHNIQUES

As mentioned earlier, subjective determinations of accomplishment should be avoided when possible. For some work they are unavoidable, however. Let us say the owner has contracted for project scheduling services. This is often called a level-of-effort contract because a certain level of effort on the part of the contractor will be expended, and this amount is not generally proportional to the "products" that result, if in fact any identifiable products exist. If the scheduling contract in question is firm fixed price (say, $50,000), partial payments are generally based on elapsed time or cost—in this case, perhaps through prearranged monthly installments, or in some proportion to the number of hours expended during the month for scheduling.

At times, one sees the use of quasi-objective techniques in conjunction with a subjective basis. In the scheduling contract, for example, the owner may ask the scheduling contractor (or consultant) to estimate the percentage of completion of the project scheduling effort. If the contractor replies that he is 40% finished with scheduling duties, a payment of 0.40 of $50,000, or $20,000, may be made. This demonstrates the "mask" of objective measurement applied to the face of subjectivity. For if one investigated the logic used by the contractor to achieve the 40% estimate of completion, it would probably have a cost or time basis. He may have decided that 1,000 hours would be used to perform all scheduling work and that 400 hours have been spent so far—thus the conclusion that 40% of the work has been accomplished. This is a cost-based determination. On the other hand, he may have assumed (or known) that the scheduling effort will take place over 10 months and that his schedulers have been working on the project for 4 months—

again, leading to the estimate of 40% completion. This logic is based on elapsed time.

## VARIATIONS AND COMBINATIONS

Of the five techniques described, there are thousands of variations and combinations that the owner or contractor may use. For example, milestone techniques may allow the determination of intermediate progress between milestones in a subjective or equivalent-units manner. And there is nothing magic about the 0–100 or 50–50 allocations. These may instead be chosen as 10–90, 25–75, or even 12–23–42–23. When this is done, however, the percentage attributed to each event (or milestone) becomes essentially the same as the weighted value applied to intermediate milestones. As owners and contractors become more familiar with each technique, they soon realize that these techniques are all simply variations on milestone methods. And they are all designed to achieve the same objectives: to identify measurable accomplishments and assign each one a relative payment value.

## SELECTING PROGRESS PAYMENT SCHEMES AND WEIGHTED VALUES

Regardless of who determines the methodology and values used to make progress payments, the determination should be made as early in the contracting process as possible. It is suggested that the owner decide on the details of progress payments prior to creation of the RFP and that bidders be made aware of the specifics of how progress is to be measured and paid for. The resulting terms should then be made part of the contract documents. In any case, they should be established prior to the measurement of progress for the first payment period.

Many owners define the methodology to be used and allow contractors to bid the weighted values themselves. Then an examination of the reasonableness of bid amounts is made part of the evaluation process, with any discrepancies isolated for preaward negotiation; again, the owner will wish to back-end load, whereas the contractor will wish to front-end load values. Other owners allow the bidders to propose payment schemes and/or relative amounts to each intermediate milestone or unit of measure. Of these two approaches, the former is recommended. The owner should consciously determine the payment scheme and weighted values; it can do this without knowing bid prices by placing a percentage of the total bid amount in the RFP for each line item.

In order to determine the payment approach, the owner should review the scope of work, identify its control objectives (such as what activity does the owner want to promote), and communicate the selection details to all bidders. Although contractors will object, as they should, to unrealistic methods or values, the owner should take the lead—agreeing to contractor-proposed changes, when they make sense, during the prebid meeting, preaward meeting, or award negotiations.

## RECOMMENDED PROCEDURE FOR PROGRESS MEASUREMENT AND PAYMENT

A recommended procedure for periodic (monthly) progress determinations and subsequent payments is depicted in Figure 17. In addition, the entire payment determination and paperwork flow is shown in Figure 18. Both of these depictions apply to objective measurements of progress, with the contractor making an initial determination subject to prearranged criteria suggested in this chapter.

**FIGURE 17.**  Contract progress payment process.

**FIGURE 18.**   Typical paperwork flow: progress payments and billings.

A basic goal of any payment procedure should be that resolutions of progress and commensurate amounts due the contractor should be made prior to receipt of contractor billings, or invoices. Time of invoice receipt should not be the time for both parties to begin deciding how much is earned and owed. A monthly "progress estimating" methodology is recommended, with the results documented on owner-provided progress estimating forms similar to those shown as Exhibits 10 and 11. These are the tools by which the owner and contractor agree and document their agreement. Once an invoice is received and the billing amount agrees with the amount previously determined through the progress estimating exercise, the invoice may be honored.

The format and payment items contained in the estimate forms should be determined during the commencement of the contract administration period and should comply with any contractual payment terms and conditions. The contract administrator furnishes the contractor with "master" payment estimate forms, and the contractor uses these to determine progress each month. It then submits completed forms to the owner (contract administrator), and agreement is evidenced by the appropriate signatures. A copy of the signed progress estimate form(s) is sent to the owner's accounting group. The contractor then prepares a monthly invoice to reflect the amount shown as the total on the progress estimate forms signed by the owner. It submits the invoice, along with the signed estimate form. The accounting (or accounts payable) group receives the invoice, verifies it for accuracy, and compares it with the separately received progress estimate forms (sent by the contract administrator). If no discrepancies are noted, the payment is made.

A major premise in this suggested process is that the accounting group is not responsible for verifying performance prior to payment, or for ensuring contractual compliance. This is the duty of the construction manager and supporting staff—contract administrator, construction specialist, and so on. A common tendency, however, is for overlapping or gaps in responsibility concerning the verification of progress. If more than one individual "signs off" on progress estimate forms, there is some risk that each one will rely on the other to check certain elements. In order to ensure that this does not occur, each person reviewing estimate forms or invoices should be aware of his or her specific responsibility in the process. As shown in Exhibit 10, each person who signs or initials the forms representing progress should be indicating that a certain and specific review has occurred. Some owners require the contractor to sign approved estimate forms as well, and this is recommended in order to avoid potential disputes.

## FINAL PAYMENTS

The suggested procedure for monthly progress payments should also be followed when the final contract payment is made, with a few ad-

**EXHIBIT 10. Summary of Contract Progress Payment Estimate**

## SUMMARY OF CONTRACT PROGRESS PAYMENT ESTIMATE

Contract No. _____

Contractor _____

Period Covered:

from _____ to _____

Sheet _____ of _____

(detailed sheets attached)

### Owner's Review and Acceptance

1. Compliance with Commercial Terms of Contract

   _____
   Contract Administrator

2. Compliance with Technical Requirements

   _____
   Supervising Engineer

3. Costs verified, checked, and balanced; escalation (if any) amounts correct

   _____
   Construction Accountant

4. Approval to honor invoice if in accordance with this summary and attached detail sheets

   _____
   Construction Manager

### Summary of Progress

| | |
|---|---|
| Cumulative total as of last period | $ _____ |
| Earned this period | $ _____ |
| Total earned to date | $ _____ |
| Less retention | $ _____ |
| Balance | $ _____ |
| Less previous payments | $ _____ |
| Less Backcharges | $ _____ |
| Less other charges | $ _____ |
| Amount to be invoiced | $ _____ |

### Contractor Attestation

I hereby certify that this Contract Progress Payment Estimate represents the work performed for the period shown above.

_____        _____
contractor's signature                    date

**EXHIBIT 11. Detailed Contract Progress Payment Estimate**

## DETAILED CONTRACT PROGRESS PAYMENT ESTIMATE

Contract No. _____
Contractor _____

Period Covered:
from _____ to _____

sheet _____ of _____
Prepared by: _____
Checked by: _____

| Payment Item | Description | Unit of Measure | Price | Earned as of Last Period | | Earned This Period | | Earned to Date | |
|---|---|---|---|---|---|---|---|---|---|
| | | | | Quantity | Amount | Quantity | Amount | Quantity | Amount |
| | | | | | | | | | |
| | | | | | | | | | |
| | | | | | | | | | |
| | | | | | | | | | |
| | | | | | | | | | |
| | | | | | | | | | |
| | | | | | | | | | |
| | | | | | | | | | |
| PAGE TOTALS: | | | | | | | | | |

181

ditional precautions. In general, a checklist should be used to ensure that all remaining performance has been completed, including final testing and receipt of any required contractor submittals. More detailed recommendations for final payments are contained in Chapter 17.


## RETENTION

A time-honored tradition in the construction industry calls for withholding of earned payment amounts until a later time, when they are released to the contractor. This practice is called retention. In general, specified amounts that have been earned through progress on the work are not paid immediately but are held by the owner as an assurance that the remaining work will be performed in accordance with the contract documents.

Retention is most commonly 10% of amounts due, with release made upon successful completion of the work. It can be determined differently, and it is becoming increasingly popular to establish retention as a certain amount rather than as a percentage (say, $100,000), or to vary the absolute amount or percentage representing retention as the work progresses. In order to discuss retention more closely, let us first examine the reasons behind this practice. The primary reason cited for using retention is to ensure contractor performance throughout the remaining duration of the contract performance period. Other reasons may include:

1. *To Motivate the Contractor to Complete the Work.* When used for this reason, retention (or its release) becomes a form of price incentive.
2. *To Cover the Risk of Latent Errors or Omissions.* Should these surface after the contractor has left the project site, they may be corrected by using the retained amount.
3. *To Encourage Contractors to Return to the Work After a Planned Demobilization.* Should the value of remaining work constitute a minor price incentive, the contractor may be encouraged to return if completion of that work is a prerequisite to release of a larger retention amount.

The release of retention should vary with the owner's control objectives and the scope of work in question. In addition to the typical release conditions, such as completion of all contract work, retention may be released upon:

1. *Partial Completion of Specified Portions of the Work.* Retention may be released in stages as the work progresses.
2. *Passage of a Specified Period of Time.* The owner may wish to wait 30, 60, or 90 days or more following completion before releasing it to the contractor.
3. *Achievement of Certain Performance.* If operating equipment, machinery, or systems are involved, the owner may make certain performance of each a prerequisite to retention release (successful operation, completion of systems testing, after commercial operation, and so on).

Whenever retention is used by owners for any of the preceding reasons, its use should be based on actual conditions and perceived risk rather than on tradition or because "we have always retained 10%—doesn't everyone?"

Contractors object to the use of retention in most cases, and rightly so, for it represents their money, the time value of which may be significant. Owners should always realize that it is they who pay this cost—not the contractor. Contractors have to pass this cost along to the owner in the form of higher bid amounts. For large contracts, retention may represent considerable cost to the owner. And owners should recognize the cost of retention and balance it against the value of the increased control they may gain through its use.

Unjustified witholding of any monies due the contractor by the owner, including retention, is a poor contracting practice to be avoided. In addition, many jurisdictions place specific restrictions on this practice. When inserting retention terms into contract documents, owners should verify that they are in accordance with any statutory provisions and are in fact enforceable.

## REPRESENTATIVE CASES

1. A lump-sum contract totaling $1.5 million for furnishing and installing a concrete chimney and steel liner was awarded for an electric generating station to Contractor B. Since a lump-sum bid was requested the project purchasing agent saw no need to ask for a price breakdown other than the total lump-sum price submitted by each bidder. At the end of Contractor B's first full month of work at the site, a progress payment meeting was held. Contractor B reported $600,000 of work accomplished. The owner's contract administrator (the purchasing agent

was back at the home office working on the next project) felt that only 10% of the chimney had been installed to date and that Contractor B therefore was due only $150,000. Bitter arguments ensued, with the contractor claiming "front-end" costs of mobilization, material purchases, and so on, that although not incorporated into the chimney were still incurred. The contract administrator, seeking a compromise, asked the contractor to prepare a schedule of values for the work and offered that once agreed upon, it would become the objective basis for partial payment. Contractor B prepared and presented the schedule of values. It showed a 90% recoup of the lump-sum price by the third month, even though the contract was to take 10 months to complete. Contractor B refused to change its position and threatened to leave the project if its schedule of values was not accepted. Some years later the contract administrator was promoted to manager of contract formation. He personally prepared the proposal form for a lump-sum RFP for chimney erection on another project. This is how he structured the bid form:

| | Item | Material | Labor |
|---|---|---|---|
| A. | Excavation for foundation | $_____ | $_____ |
| B. | Concrete foundation, complete | $_____ | $_____ |
| C. | Concrete chimney to elevation 200' | $_____ | $_____ |
| D. | Concrete chimney from elev. 200' to elev. 400' | $_____ | $_____ |
| E. | Concrete chimney from elev. 400' to elev. 600' | $_____ | $_____ |
| F. | Concrete chimney from elev. 600' to elev. 800' | $_____ | $_____ |
| G. | Steel liner to elev. 400' | $_____ | $_____ |
| H. | Steel liner from elev. 400' to elev. 800' | $_____ | $_____ |
| I. | Ductwork, complete | $_____ | $_____ |
| J. | Aircraft warning lighting system, complete | $_____ | $_____ |
| K. | Chimney painting, complete | $_____ | $_____ |
| L. | Lightning protection system | $_____ | $_____ |
| | Total material | $_____ | |
| | Total labor | | $_____ |
| | Total lump-sum price, including items A–L above | $_____ | |

A master progress estimate sheet was prepared containing Items A through L, and payment was made as each item was completed. During the course of construction, painting of the chimney was deleted from the scope of work. In order to make an equitable adjustment in the lump-sum price, the bid (and contract) amount for Item K was used to credit the owner for the deletion. This avoided costly and troublesome negotiation over the value of the deleted work.

2.  A large pressure vessel for a chemical plant was ordered on a furnish-and-install contract using a lump-sum price structure with provisions for escalation in material cost. The successful bidder quoted a price of $2 million, to be paid in 20 equal installments (it was to take 20 months for fabrication, shipment, installation, and testing). The owner's contract administrator realized that if this was accepted, the company would be paying on the basis of elapsed calendar time with no regard for the actual progress of the contractor. During the preaward meeting, he proposed a schedule of values that tied the payment of specific amounts to accomplishment of specific progress milestones. The contractor, wanting the work, agreed. Progress payment forms were created as follows:

| | |
|---|---|
| Delivery of new materials at contractor's fabrication plant | $ _____ |
| Completion of vessel walls | $ _____ |
| Completion of vessel ends (caps) | $ _____ |
| Shipment of vessel and auxiliaries from plant | $ _____ |
| Receipt of vessel and auxiliaries at project site | $ _____ |
| Initial setting of vessel on foundation | $ _____ |
| Installation of auxiliaries | $ _____ |
| Final installation of vessel | $ _____ |
| Completion of Hydrostatic Test | $ _____ |
| Final acceptance of vessel | $ _____ |
| Total lump-sum price | $ _____ |

Upon completion of fabrication, the contractor's truck drivers struck for eight months and halted all shipments. No progress payments were made by the owner. Had the contractor's proposed time-oriented payment terms been accepted, a total of eight payments ($800,000) needlessly would have been made, with no progress accomplished during that period.

3.  A unit-price contract for furnishing and placing concrete for foundations and elevated decks was awarded Contractor C. At the end of the first month, Contractor C's records, based on invoices from the concrete batch plant it was using, showed 1,500 cubic yards poured. At a contract price of $40 per cubic yard, it requested payment of

$60,000. Suspicious of this tremendous progress, the owner's contract administrator asked the project civil superintendent to calculate the amount of concrete placed by visually inspecting the foundations and reviewing the drawings. From this check the superintendent determined that 900 cubic yards had been placed. Contractor C was told to invoice $36,000 for concrete in place. Furious, the contractor's project manager investigated his operations and found the following:

1.  300 cubic yards had been sold by the batch plant to Contractor C but had been used for another nearby project.
2.  200 cubic yards erroneously had been invoiced by the batch plant to Contractor C's account but actually had been received by another contractor.
3.  100 cubic yards had been spoiled by Contractor C's concrete trucks before ever arriving at the owner's construction site.

    4.  In order to ensure that a large cooling tower passed final commercial testing, the design engineer inserted a clause into the specifications calling for a 20% retention until completion of the testing. As the contract sum amounted to $35 million, and preoperational testing was not to commence for the project until two years after completion of the tower, all seven bidders bitterly protested this burden at the prebid meeting. They claimed, rightfully, that the final retention costs would create interest expense for the contractor (and would be passed on to the owner) of 20% of 35,000,000 × 14% interest compounded annually for two years, or $2,097,200. Cooler heads prevailed, and a graduated retention plan was substituted. The new plan called for 10% retention during the first year, reduced to 5% during the remainder of the work, and never to exceed $400,000. At the completion of work all payments due the contractor would be made, with the exception of $200,000, which would be retained until preoperational testing was complete and the tower accepted for use.

    5.  At times, retention is used by the owner as a lever not to ensure that work is performed in accordance with the contract but to ensure that a contractor who has a scheduled break in performance will actually return to the project site and complete the remainder of the work. In such a case, the schedule of values for a structural steel erection contract allowed $12,000 for erection of steel to be left out in a wall for a turbine building. The contractor was to demobilize for six months and return to complete the large wall penetration after the turbine generator was installed (the opening was left to allow the large equipment to be moved into the building). Since the total steel contract was for approximately

$4 million, the project owners felt that the $12,000 task might not be sufficient to ensure prompt return by the contractor in time to close the turbine building before the onset of winter. For that reason, a retention of $350,000 was withheld after all work except the penetration closure was completed. Obviously, the retention far exceeded the value of the remaining work, but its use did cause the contractor to return and complete the work as soon as requested. This approach cannot be implemented whenever the owner desires; it must be conceived prior to bidding and described in the RFP and contract documents. To impose retention without contractual authority is bad business practice and could be contrary to the law.

# CHAPTER 13

# NOTIFICATIONS AND CHANGE ORDERS

Nothing is more constant than change during the course of a construction project. Despite the best efforts of all concerned during the planning, formation, and administration phases of contracting, changes will almost certainly occur. This chapter describes the typical causes of contractual changes and presents a process of accommodating changes within the framework of owner–contractor agreements.

## TYPES OF CONTRACTUAL CHANGES

There are thousands of changes that take place during the design and construction of a new facility. Among these are changes to the design, changes to the schedule, price and cost changes, resequencing of design and construction activities, material substitutions, and modifications to construction methods, to name only a few. Many of these changes may be accommodated with little problem. In fact, many represent improvements to original plans or refinements made to improve performance when unforeseen events occur or when more information is available concerning anticipated events. In this chapter we will restrict our discussion to those changes that affect contractual agreements: contractual changes. These represent the most damaging of all.

Most construction practitioners categorize contractual changes into two groups: informal and formal changes. Formal changes are defined as those that are made by the owner and result in written directives to the contractor to change the scope of work, time of performance, or both once the contract is awarded. They may be thought of as conscious decisions on the part of the owner to modify the contractor's work.

The most common types of formal changes concern alterations to the design of the facility and are manifested in revisions to the construction drawings or specifications. In addition, owners often change their schedule needs or resequence the planned work of contractors to accommodate changes in the owners' needs or performance or in those of other contractors or suppliers. Formal changes have their authority in the changes clause of most contracts. This clause usually gives the owner the unilateral right to change the work and to require the contractor to comply with such changes. The additional compensation, in terms of time, money, or both, necessitated by the change, is determined in a structured, prearranged manner according to the terms of the original agreement (see discussion of pricing change orders in Chapter 3). The written directive ordering the contractor to make the change and specifying additional compensation (if any) is called a change order.

Because formal changes are generally identified before they are incorporated, based on a planned and deliberate choice by the owner, and documented by a formal modification instrument—change order—they are usually easy to handle and accommodate. Strict controls are required to assure the owner that these types of changes are identified as early as possible, thoroughly evaluated in terms of cost, schedule, and technical impact on the work, priced economically, and enacted in accordance with the contract provisions. Although the number and frequency of formal change orders is sometimes staggering, most contract administration programs are able to accommodate them with a minimum level of difficulty. The same cannot be said for the other category of change: informal changes.

Informal changes are sometimes called "constructive changes," in a legal sense. They represent modifications to the contractor's scope of work or method of performance that result from: (1) acts or omissions of the owner; (2) acts or omissions of third parties, such as other contractors, material suppliers, or God; or (3) other forces beyond the contractor's control. They are the most difficult to identify and to control. They cause the contractor to perform work differently or to perform work different from that which is required by the contract or which could be normally anticipated. The result is the same as if a formal change order had been issued. Once informal changes are identified and an agreeable resolution between owner and contractor is achieved, they should be transformed into formal change orders.

Major difficulties presented by informal changes arise because they originate everywhere and are often identified after the fact and their cost and schedule impacts are difficult to quantify. In addition, informal changes detract from the performance of unchanged work. Many times these types of changes are therefore difficult to resolve and fester into troublesome claims.

## TYPICAL CAUSES OF CHANGE

There are countless causes of change that may impact a contractor's work on such a complex, highly interdependent, and time-sensitive effort as a large construction project. Students of these phenomena, however, suggest that a large percentage of change is caused by one or more of the reasons described in the following paragraphs.

1. *Defective or Incomplete Design Information.* Owners dissatisfied with the products of their engineering departments or the performance of their design consultants would probably vote for this as a major cause. Symptoms of this problem are numerous or massive changes, or revisions, to the design drawings and specifications. Revisions to design—made when the contractor has already commenced work—are not bid competitively. This situation results in a vicious cycle: revisions to one system, structure, and so on, causing revisions to interrelated systems, structures, and so on. It is a common phenomenon on fast-tracked projects, where design is often only one step ahead (or behind) construction.

2. *Late or Defective Owner-Furnished Material and Equipment.* Many owners purchase some equipment or material and furnish it to the construction contractor(s) for erection or installation. Owners who decide to save money or exercise direct control by buying these items themselves must consider the additional responsibility and liability they assume by doing so. The decision to buy directly should be based upon an understanding of owner and contractor purchasing strengths and controls, as well as the economies involved. The question of procurement responsibility should, therefore, hinge not only upon anticipated price but also on a determination of which party is best equipped to handle the additional responsibility—who can ensure that the materials or equipment will arrive at the site on time and in the proper condition.

3. *Changes in Regulatory Requirements.* Changes in regulatory requirements, particularly for those projects that span several years or concern highly complex or regulation-prone efforts, such as nuclear power plants, represent major causes of change, both formal and informal. There is little an owner or contractor can do about these types of changes. At best, every effort should be made to anticipate changes

of this nature and to provide for their impacts in project budgets and contractual terms and conditions.

4. *Changed or Unknown Site Conditions.* Subsurface conditions represent a classic example of site conditions that may not be known or that may change during the course of contractual performance. The occurrence of subsurface water, rock, or other material often directly affects a contractor's performance. Other site changes, though not as common, also occur. These include weather, access, congestion in the workplace, and operating space, such as material laydown areas, warehouse storage, and security restrictions.

5. *The Impact of Collateral Work by Others.* When actions of a contractor interfere with, delay, or cause hardship on other contractors, informal changes may result. This risk is intensified in direct proportion to the number, proximity, and interdependency of contractors at the site.

6. *Ambiguous Contract Language and Contract Interpretations.* Extensive contract formation controls, consistent and clear documents, and contract administration training and procedures help prevent changes due to this cause.

7. *Restrictions in Work Method.* There are countless examples of owners restricting or changing contractor work methods. Unless it is specified otherwise in the contract documents, contractors are generally free to construct in the manner they deem suitable. Owners who impose restrictions on acceptable or traditional methods, absent any specific restrictions in the contract, interfere with the contractor's performance. As owners become increasingly involved with the daily operations of contractors—crossing the fine line between performance monitoring and supervision—interference becomes possible. As independent contractors responsible for the means and methods employed to satisfy the contract requirements, contractors will claim informal changes if this occurs.

8. *Late or Inadequate Contractual Compliance on the Part of the Owner.* Construction contracts are a two-way street, requiring compliance by both parties. Although much has been said regarding controls over contractor compliance, the same need exists to ensure that owners fulfill their contractual obligations. Many owners feel that their obligations are restricted to payment of contractor charges. This is not true. Virtually every contract requires (expressly or implicitly) actions on the part of the owner in addition to payment of the bill. Most of these deal with prompt and timely owner inspection and acceptance of the work product as it progresses. Owners are often required to

review design or shop drawings produced by the contractor in a timely manner. This requirement is sometimes stated as a duty that must be performed within so many days of receipt of these and other contractor submittals. It is not uncommon for owners, in their zeal to manage or monitor contractor performance, to overlook commensurate performance required on their part. The capability of the owner to meet these contractual commitments should be analyzed before they are inserted into contract documents.

9. *Delay or Acceleration.* When a contractor's work is delayed, by the owner, other contractors, or third parties, it incurs additional cost. In most instances, the contractor also is unable to meet existing contractual time commitments. The same is true when its work is accelerated, or compressed. Typical costs for delays include standby and remobilization expenses, loss of efficiency, and changes in the weather or other conditions. Acceleration costs often involve the increased charges incurred for shift time and overtime, additional work forces and equipment, and loss of efficiency. Both acceleration and delay situations illustrate that a combination of causes and effects, which in turn become causes for further effects, is often found among constructive or informal changes. Separating each, and quantifying their respective impacts in terms of time and money, is very difficult. This is one reason why timely identification and resolution of changed conditions is critical. The alternative is often litigation or prolonged negotiation.

## THE CHANGED WORK PROCESS

The truest test of any contract management program is its ability to respond to and accommodate changes. Whether they are formal or informal changes, contract administration controls and procedures need to allow the owner to perform each of the steps described in the following paragraphs in the changed work process.

1. *Identification.* It is surprising how many changes, particularly informal ones, go unrecognized until it is too late to control their impact. Changes should be identified as such as soon as possible in order to allow a thorough evaluation of their impact and a conscious, well-informed decision as to whether and how they will be enacted. Planned or potential changes, when initiated by the owner, should be transmitted to the contractor immediately. This should be done through use of a form letter similar to the notification instrument shown as Exhibit 12. As the term implies, *notifications* are merely instruments by which the owner notifies a contractor of a planned or potential change. They

**EXHIBIT 12.   Notification**

NOTIFICATION

Date: _____

Notification

Number: _____

To: _____          Contract No. _____

_____          Contract _____

_____

The enclosed ☐ information ☐ document(s) ☐ drawing(s)
☐ are proposed ☐ have been approved
for use on the above contract work.

| *No.* | *Description* | *Identification* | *Remarks* |
|---|---|---|---|
|  |  |  |  |
|  |  |  |  |
|  |  |  |  |

(attach continuation sheets if necessary)

Please review the enclosed and proceed in accordance with the instructions indicated below.

INSTRUCTIONS

☐  No change required in contract, proceed with affected work.
☐  For information only.
☐  Change in contract work, submit proposal, including all impacts on cost and time of performance, as requested by _____.
(date)

    ☐  lump sum. Do not proceed with work.
    ☐  contract unit prices. Submit estimated quantities and proceed with work.
    ☐  unit prices not included in contract. Submit prices and estimated quantities. Do not proceed with work.
    ☐  cost plus in accordance with contract provisions. Submit estimated manhours and material costs. Do not proceed with work.
    ☐  other: _____

Issued by: _____ date _____
Receipt acknowledged by: _____ date _____
Return one copy of this form, with receipt acknowledged by signature of contractor, to: _____ by _____.
(date)

_____ continuation sheets are attached.

also serve as a request for the contractor to determine the impacts on time and schedule, if any, that the change would cause and to quote on both to the owner by a specified time. Notifications (sometimes called "bulletins") differ from change orders in that they do not necessarily instruct the contractor to perform the changed work. They merely request the contractor's reaction to proposed changes. When designed and used properly, they should:

Inform the contractor of all details of the proposed change. If revised drawings or specifications are involved, these should be listed and attached.

Indicate whether the information constitutes a change, or is merely additional or promised information that does not affect change or does not deviate from the original agreement.

Request a contractor quotation, in a specified pricing manner (lump sum, unit prices, and so on) by a specified date.

Contain a clear statement as to whether the work should commence immediately or the contractor should ignore the change until a change order is issued.

Be signed and dated by the owner. Numbering notifications is also suggested for ease of reference.

Require the contractor to acknowledge receipt of the notification by returning a signed copy thereof. This indicates not that the contractor agrees to perform the affected work but merely that the contractor is aware that the change has been proposed.

The notification process fits situations where changes are proposed by the owner. For this reason, it is well suited for most formal changes. However, many times changes are proposed or identified by the contractor. This is often the case with the other category of changes—informal or constructive changes. When this occurs, contractors should notify the owner of the change and its impact, even if the changed work has been completed. This is done through correspondence from the contractor to the owner identifying the change and describing the actual or anticipated impact on cost, schedule, or technical performance. Contractor identification of this nature is generally in the form of a change request or a claim. Like a notification, a change request or claim satisfies the requirement that changes be identified and communicated to the other contractual party.

2. *Evaluation.* Once a change has been identified, by either the owner or contractor, the owner must make a determination as to whether

it wishes to adopt the change. If identification occurs after the fact, the impact of the change must be estimated or determined. The evaluation of a proposed or enacted change involves a review of the contractor's quote (or claim) for accuracy, compliance with any contractual provisions, and reasonableness. Just as fair price estimates help owners to evaluate and judge original contract bids, so do they help evaluate change quotations. They should be considered for this purpose. The contract administrator requests such an estimate on a form similar to the "Request for a Fair Price Estimate," Exhibit 13. Just as the contractor is instructed as to the required pricing method, so should this request

---

**EXHIBIT 13.   Request for Fair Price Estimate**

### REQUEST FOR FAIR PRICE ESTIMATE

TO: _____          Contract No. _____

FROM: _____          Contract: _____

Please prepare a FAIR PRICE estimate based on ☐ the attached, or
☐ Notification No. _____

and return by _____.
                    (due date)

**Required Pricing Method:**

☐  Lump Sum
☐  Unit Prices (with extensions)
☐  Cost Plus (with estimated cost)
☐  Contract Unit Prices ☐ Apply, ☐ Do Not Apply
☐  Contract Cost Plus Markups ☐ Apply, ☐ Do Not Apply

Additional Instructions: _____

_____

_____

Attachments: _____

_____

_____

_____

_____

                                              (contract administrator)

be made to those responsible for preparing the fair price estimate. Should pricing method and format between these two differ, evaluation is hampered.

3. *Approval.*   Often owners go back to contractors requesting additional information or rejecting quotations and asking for a revised quote. Sometimes negotiation is required. When both parties have agreed to the need and amount of additional compensation, a change order is issued. The documentation and approval of the change identification and evaluation process is done through a form similar to the change order approval form shown as Exhibit 14. Owner management indicates its approval or rejection of the proposed (completed) change and any pricing and schedule changes. This form serves to document the evaluation and to justify the incorporation of the change into the contractual agreement.

---

**EXHIBIT 14.   Change Order Approval Form**

CHANGE ORDER APPROVAL FORM

Number _____

Contract No. _____

Source of Change                    Date

Contract: _____

- ☐ Notification No. _____          _____
- ☐ Contractor Claim                   _____
- ☐ Contractor Request                 _____
- ☐ Other _____                 _____

Pricing Method to be Used: _____
Estimated (or Lump-Sum) Price: _____

Cost Evaluation

| Item No. | Description | Pricing Method | Quantity | Fair Price Estimate | Contractor Quote |
|----------|-------------|----------------|----------|---------------------|------------------|
|          |             |                |          |                     |                  |
|          |             |                |          |                     |                  |
|          |             |                |          |                     |                  |
|          |             |                |          |                     |                  |
|          |             |                |          |                     |                  |
|          |             |                |          |                     |                  |
|          |             |                |          |                     |                  |
|          |             |                |          |                     |                  |
|          |             |                |          |                     |                  |

Schedule or other impacts: _____

_____

Prepared by: _____          _____
                    (contract administrator)              (date)

Reviewed by: _____          _____
                    (construction specialist)             (date)

☐ Approved ☐ Rejected by: _____     _____
                                    (construction manager)          (date)

If approved, list Change Order number(s) to be issued: _____

                                                      _____

                                                      _____

Additional Disposition Instructions: _____

_____

_____

_____

---

4. *Incorporation.*   Once a change order is approved by the owner,
it is issued to the contractor to modify the terms of the original agreement.
It may incorporate the changes of one claim, notification, or contractor
change request or, as is often the case, consolidate the collective results
of several of these identification instruments. Change orders should be
numbered and dated and refer to the changes clause of the contract
documents. They should thoroughly describe the change and the revision
to contract price or time of performance necessitated by that change.
In addition, change orders should also be signed by the contractor,
indicating his acceptance of the changed work as well as the change
in compensation. An example change order form is shown as Exhibit
15.

---

**EXHIBIT 15.   Change Order**

CHANGE ORDER

Contract No. _____          Change Order No. _____

Contract _____              Project _____

To: _____

    _____

    _____

*(continued on page 198)*

**EXHIBIT 15**  *(continued)*

In accordance with Paragraph _____ of the General Conditions of the above named contract, the following change is made and incorporated into said contract:

_____

_____

_____

_____

The Contract Price is changed as follows: _____

_____

The Contract Performance Dates and/or Durations are changed as follows: _

_____

All other terms and conditions of subject contract remain in full force and effect.

Prepared by: _____    _____
                        (contract administrator)                (date)

Approved by: _____    _____
                        (construction manager)                (date)

Accepted by: _____    _____
                              (contractor)                      (date)

Please sign and return one copy to: _____

FOR OWNER USE ONLY

Original Contract Price:            _____
Current Contract Price:            _____
Notification No.                      _____
Change Order Approval
Form No.                               _____

5. *Payment.*   Payment for change order work should follow the same procedure as described for normal contract progress payments. The exception is that changed work should be identified and listed separately on progress estimates and invoices. It is strongly recommended that original and changed work not be commingled during this or any other step of the changed work process.

The preceding steps are presented in graphical form in Figure 19. This figure shows the typical participants responsible for the process as well

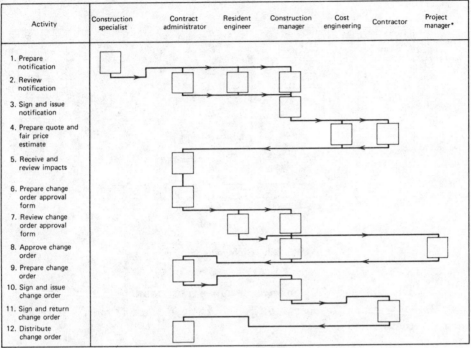

| Activity | Construction specialist | Contract administrator | Resident engineer | Construction manager | Cost engineering | Contractor | Project manager* |
|---|---|---|---|---|---|---|---|
| 1. Prepare notification | | | | | | | |
| 2. Review notification | | | | | | | |
| 3. Sign and issue notification | | | | | | | |
| 4. Prepare quote and fair price estimate | | | | | | | |
| 5. Receive and review impacts | | | | | | | |
| 6. Prepare change order approval form | | | | | | | |
| 7. Review change order approval form | | | | | | | |
| 8. Approve change order | | | | | | | |
| 9. Prepare change order | | | | | | | |
| 10. Sign and issue change order | | | | | | | |
| 11. Sign and return change order | | | | | | | |
| 12. Distribute change order | | | | | | | |

* May be required for certain change orders

**FIGURE 19.**  Change order process.

as the general flow of paperwork involved. Because the control and documentation of changes represents one of the most critical challenges during the performance period, and many owner programs are weak in this area, the use of carefully designed documents and records during this process deserves particular attention.

## CHANGED WORK DOCUMENTS

Generic examples of documents used to control and record the changed work process are included as exhibits in this chapter and shown in relation to one another in Figure 20. Each of the five steps described earlier is represented by a specific document or documents, and their recommended order of occurrence is indicated. Because so many changes commonly occur, and considering that many changes often revise work or conditions that have been previously changed, a disciplined, structured approach to this process is essential. Contract files should fully document the changed work process. When this occurs, two primary benefits are realized:

**FIGURE 20.** Changed work documents.

1. The changed work process becomes more manageable.
2. Operational or cost auditing of changed work is more easily accomplished.

The changed work process described in this chapter, together with examples of documents used and depicted in Figure 20, meet both of

**EXHIBIT 16.   Notification Log**

NOTIFICATION LOG

Page _____

Project: _____
Owner: _____

| Notification Number | Contract Number | Contractor | Date Issued | Date Receipt Acknowledged | Remarks |
|---|---|---|---|---|---|
| | | | | | |
| | | | | | |
| | | | | | |
| | | | | | |
| | | | | | |

**EXHIBIT 17.  Change Order Log**

CHANGE ORDER LOG

Page _____

Project: _____    Contract No. _____

Owner: _____    Contractor: _____

| Change Order Number | Description | Notification Number | Pricing Method | Estimated or Lump-Sum Price | Date Issued | Date Copy Received |
|---|---|---|---|---|---|---|
|  |  |  |  |  |  |  |
|  |  |  |  |  |  |  |
|  |  |  |  |  |  |  |
|  |  |  |  |  |  |  |
|  |  |  |  |  |  |  |
|  |  |  |  |  |  |  |
|  |  |  |  |  |  |  |
|  |  |  |  |  |  |  |
|  |  |  |  |  |  |  |

these objectives. The audit trail provided by a complete "set" of documents used during each step is clear and easy to follow. Many features of each document serve this purpose. A review of each exhibit shows how, when practical, each refers on its face to the number or other identification of associated documents that precede or follow. Without these ties, or without the documents themselves, control and review of changed work becomes an unbearable task. Not only should each changed work incidence be carefully controlled and documented, but the cumulative effect of changes should also be reported and analyzed by owner management. This type of effort is more fully described in Chapter 19.

Other compliance controls should be used by the contract administrator to ensure that changes are being handled properly. For example, he or she should consider the use of a notification log to record and track the status of notifications. An example is shown as Exhibit 16. In addition, a log of all issued change orders is recommended. A change order log, such as that shown in Exhibit 17, and a changed work history form, such as that shown in Exhibit 18, help the contract administrator identify and track the status of issued change orders. They are extremely useful when the amount of changes increases and as checks against change orders identified on progress estimates and/or contractor invoices.

**EXHIBIT 18.  Changed Work History**

## CHANGED WORK HISTORY

CONTRACT NO. _____

CONTRACTOR _____

| Notifi-cation No. | Description | Issued Date | | Ac-know. | Proposal/Estimate Submitted | | | | Contrac-tor's | Proposal/Estimate Resubmitted | | | | Re-jected | Resolved | | | Final Cost | Comments |
|---|---|---|---|---|---|---|---|---|---|---|---|---|---|---|---|---|---|---|---|
| | | To Contr. | To Est. | Recv'd | Contractor | | Estimating | | Quote | Contractor | | Estimating | | | No. | Change Order | | | |
| | | | | | Sched. | Act. | Sched. | Act. | | Sched. | Act. | Sched. | Act. | | | Issued | Accept. | | |
| | | | | | | | | | | | | | | | | | | | |
| | | | | | | | | | | | | | | | | | | | |
| | | | | | | | | | | | | | | | | | | | |
| | | | | | | | | | | | | | | | | | | | |
| | | | | | | | | | | | | | | | | | | | |
| | | | | | | | | | | | | | | | | | | | |
| | | | | | | | | | | | | | | | | | | | |
| | | | | | | | | | | | | | | | | | | | |
| | | | | | | | | | | | | | | | | | | | |
| | | | | | | | | | | | | | | | | | | | |
| | | | | | | | | | | | | | | | | | | | |
| | | | | | | | | | | | | | | | | | | | |

## EXCEPTIONS TO THE RECOMMENDED CHANGED WORK PROCESS

Occasionally (or all too frequently), circumstances arise that necessitate changed work but do not allow the time for a formal process such as described in this chapter. Emergencies or other unusual conditions must be accommodated through more expedient means, and often these involve changed work. When this occurs, contractors are usually instructed (sometimes verbally) to proceed with the changed or affected work immediately, with resolution of its costs at a later time.

Those responsible for managing construction efforts on the part of the owner should have the authority to accommodate such emergencies or unforeseen conditions as rapidly as possible. By the same token, however, care should be taken to ensure that this exception does not become the rule, that emergency authorizations are not used excessively or to implement changed work that could be processed under more controlled methods. Simply because standard change-control processes cannot be implemented for emergency work does not mean that all control should be abandoned. When this work takes place, it is usually performed in a cost-reimbursable manner. This is understandable, as the scope and cost of such work is usually impossible to quantify before proceeding. However, the same types of controls recommended for cost-reimbursable contracting should be implemented. Some owners issue "emergency work orders," or other change-oriented instruments to cover this type of occurrence. Time sheets, material receipts, and other evidences of cost should also be collected on a daily basis. When the work is completed, or proceeds to the point where either the emergency is past or a reasonable estimate of the cost may be established, the issuance of a standard change order is recommended. In this fashion, *all* changed work for a contract is eventually documented and controlled in essentially the same manner. Because changed work represents one of the primary areas of commercial risk, it is dealt with in other chapters as well. These include Chapters 15, 19, and 20.

## REPRESENTATIVE CASES

1. A unit-price contract for furnishing and installing structural and miscellaneous steel contained an extensive price list for virtually every category of item that could be needed for a coal-fired power plant project. Of the several hundred priced items, one was for furnishing and installing "metal building frames and louvers" contemplated for the turbine building air exhaust system at a price of $40 per pound of

steel. After 18 months of work, the design was changed to require the addition of eight "exhaust vents with adjustable vanes." A change order was issued upon a lump-sum quotation from the contractor of $88,000. While reviewing the change order files, an independent auditor later pointed out that the additional "exhaust vents with adjustable vanes" were identical to the "metal building frames and louvers" for which unit prices existed in the contract. Since each vent weighed approximately 200 pounds, the use of the existing unit prices would have cost the owner $8 \times 200 \times \$40$, or $64,000, rather than the $88,000 authorized by the change order.

2.   In order to maintain budget control, the project manager adopted a single limit of $25,000 for individual change orders issued in the field by the on-site construction manager. A copy of each field-issued change order was reviewed by the home office to ensure that this policy was followed. A complete review of all change orders took place after the project was completed and all contracts closed out. At that time, it was discovered that the site manager never issued individual change orders exceeding $25,000 for the same work. For example, the rerouting of buried pipeline called for $72,000 in extra work. The construction manager divided the work into three separate change orders, each issued several days apart and interspersed with other nonrelated change orders. The change orders in question were:

1.   C.O. No. 109, reroute buried piping, $24,500, issued July 19
2.   C.O. No. 113, reroute buried piping, $24,000, issued July 23
3.   C.O. No. 121, reroute buried piping, $23,500, issued July 26

The construction manager had in effect issued a $72,000 change order in 3 installments all within one week—contrary to the spirit, if not letter, of the project manager's policy.

3.   A lump-sum contract for structural steel erection contained unit prices for potential additions or deletions to the scope of work. For a particular size of wide flanged steel members (W12 × 115) the add price was $1.15 per pound, and the delete price was $0.85 per pound. Before the contractor ever began work on building no. 2, a flood of drawing revisions was issued. Among many other changes, the drawings alternately deleted and added steel members of the size W12 × 15 as follows:

Revision Batch No. 1: delete 985 pounds W12 × 15 (March 15)
Revision Batch No. 3: add 1,785 pounds W12 × 15 (April 30)

Revision Batch No. 4: delete 640 pounds W12 × 15 (May 28)
Revision Batch No. 7: add 1,800 pounds W12 × 15 (July 14)

A change order was issued following each notification to adjust the lump-sum price according to the additions or deletions in question. The cumulative effect of the four change orders in question was $5,504:

Delete 985 pounds (0.85)   =  $   837.25
Add 1,785 pounds (1.15)   =  $2,052.75
Delete 640 pounds (0.85)   =  $   544.00
Add 1,800 pounds (1.15)   =  $2,070.00
       Total                         $5,504.00

In actuality, these were merely "paper" changes, as no work had begun on the affected areas. The contractor was overcompensated because the *net* effect of the additions and deletions (be it a reduction or increase in quantity) is the only basis for an adjustment in contract price. In this case, the net effect was:

(985)
1,785
(640)
1,800
1,960 pounds net addition

And the owner should have issued one change order covering this net addition in the amount of 1,960 × 1.15 = $2,254, rather than the $5,504 authorized by the four separate change orders.

4.   The owner of a petroleum tank farm some 2,000 miles from its home office, where design and project management took place, implemented a policy calling for the use of two distinct change order types: (1) home office change orders issued for changes due to revisions to home office–issued drawings and specifications; and (2) field office change orders issued at the construction site for site-related changes. To ensure that change orders were not issued by both sources to cover the same activity, procedure called for duplicate copies of each field change order to be sent to the contracts department at the home office. The procedure served its purpose well, until the field contract administrator fell behind in his copying and mailing chores and became lax in describing the cause and nature of the work on the face of field change order forms.

As a result, the home office was not adequately informed of the field orders and consequently issued change orders of its own, resulting in duplicate or overlapping payments to the contractor that went undiscovered. One of these instances involved the relocation of underground utilities. The home office realized from existing drawings of an adjacent plant that gas, water, and electrical lines would have to be relocated to make room for a new tank foundation. The quoted price and change order included appropriate compensation. By the time the home office change order had been issued, however, the contractor had begun the relocation work and had requested and received a field change order for the relocation from the contract administrator, who assumed the underground utilities were a site-generated occurrence. His description of the work stated simply "extra labor and material for tank foundation installation." Not realizing this actually meant utility relocation, the duplication went unnoticed in the home office when they finally received a copy of the field change order months later.

5.   In order to expedite the relocation of the main access road to a pulp and paper mill under construction (floods had destroyed the existing bridge approaching the site), the owner's contract administrator issued a cost-plus change order to the general contractor based upon a hasty estimate of $50,000. A copy of the order was sent to the home office accounting section, and the next invoice from the contractor included the change order number, date, and $50,000 as the amount. It was paid and forgotten. As it turned out, the actual cost to the contractor amounted to $26,500, and with a contractual markup on changed work of 20%, the change order should have been rescinded by one based on actual costs as soon as they could be ascertained (in this case, $31,800). Had a cost-plus work order been issued for an estimated amount of $50,000 and the accounting section instructed not to honor invoices based on any other instrument than change orders, this overpayment of $18,200 would have been avoided.

# CHAPTER 14

# BACKCHARGES

During the course of a project, it may become necessary for the owner to perform certain work operations, or have these work operations performed by others, in instances where a contractor or supplier for various reasons is unable or unwilling to perform the work operation determined to be its responsibility and included under the scope of work of the contract. When these situations occur, the cost of performing the work should be charged to the offending contractor or supplier in the form of a backcharge.

## COST COLLECTION

Other than administrative time and expense, the net cost effect on the owner resulting from a backcharge incident should be zero. That is, the additional cost to the owner for having the work performed by another contractor (or doing it itself) should equal the deduction from payments due the offending contractor. Some owners pay for the additional work and later invoice the offending contractor for reimbursement. This is not recommended, as it places the owner in a defensive position—the burden (and risk) of collection is unnecessarily placed on the owner. To avoid this, it is suggested that the owner deduct the backcharge amount from payments due the contractor by either a negative entry on monthly progress estimates or a deduction (credit) to the contractor's invoice. In cases where the backcharge situation is discovered only after the offending contractor has left the site and been paid in full, the owner has little choice but to invoice him and pursue collection. Whether to add a markup to the backcharge amount to cover the owner's "handling fee" is up to the individual owner. Usually, however, the nuisance value and ill feelings this creates far outweigh the minor amount to be charged.

## BACKCHARGE SITUATIONS

In order to understand the role of backcharging in a construction environment, let us examine some typical situations where a backcharge is in order.

### Example 1

Contractor A has been awarded a contract for excavation and pouring concrete building foundations. Included in the scope of work are placing and installing steel anchor bolts in the base slab of a large structural steel building. When this slab was poured, the contractor neglected to include certain bolts that were added by a recent drawing revision and are required to secure several major structural steel column base plates. The structural steel erection contractor (Contractor B) arrives at the site and discovers the omission immediately. Contractor A has demobilized and left the site and will not return for several months to work on other foundations. Contractor B, therefore, is issued a change order to correct the deficiency. Contractor A should be backcharged the amount of that change order.

### Example 2

Contractor C, a piping installation contractor, is receiving and unloading a shipment of owner-furnished prefabricated pipe when it realizes that extensive weld preparation of the ends of each spool piece has not been performed. The weld preparation is specified by the purchase order for pipe from Vendor D. Vendor D is notified, admits the error, and suggests that the pipe be shipped back to his plant and corrected. He expects to have the pipe returned to the site in six months. Rather than delay Contractor C for this period, Vendor D agrees to pay for the rework if done in the field by Contractor C. Vendor D signs a backcharge form, a change order is issued to Contractor C to perform the weld preparation, and the amount is deducted from Vendor D's billing.

### Example 3

An employee of Contractor E, suffering from an extreme hangover, mistakenly drives a bulldozer over a carton of sensitive (and fragile) electrical instruments waiting to be installed by an electrical contractor, F. Contractor F asserts that he has no contract or legal relationship with Contractor E and demands a change order from the owner covering the replacement cost of the damaged items. Contractor E agrees to be backcharged for the amount in question.

## Example 4

While starting up a large pump in order to test a completed fuel oil system, the owner hears a piercing sound and immediately shuts the motor off. On inspecting the motor, he finds the piping contractor (Contractor C) carelessly left a bundle of welding rods inside the pump's intake piping. The pump is destroyed and a replacement ordered. Contractor C has long ago left the site and received final payment for its contract. A mechanical subcontractor on site agrees to install the replacement pump, and a backcharge to Contractor C for the cost of the pump as well as the installation work is issued. An invoice for the total cost is sent to Contractor C.

## SUGGESTED BACKCHARGE PROCEDURE

The contract administrator should be notified of any backcharge work prior to commencement of such work. The administrator determines the scope of the work to be performed based on discussions with project personnel, correspondence, photographs, and so on, and requests an estimate from the party who will be performing the backcharge work. The cost estimate may be submitted on the basis of a lump-sum price, unit prices, including estimated quantities, or cost plus with an estimated total cost.

A backcharge form (Exhibit 19) is prepared by the administrator and contains the scope of the work to be performed and total estimated (or actual) cost. The backcharge form, including any applicable back-up information such as details of the cost estimate, correspondence, and telephone or meeting memoranda, is forwarded to the offending contractor or supplier (for whom the work is to be performed) for signing. This signature should be obtained prior to commencement of the backcharge work if possible. If the offending contractor is a material or equipment vendor, its representative is immediately called to the site to inspect the product and sign the backcharge form. Copies of the backcharge form are distributed to appropriate project personnel, including accounting, who are responsible for invoicing the contractor or supplier upon completion of the backcharge work (or periodically,

**EXHIBIT 19. Backcharge**

BACKCHARGE

Contract/Purchase Order Number _____   Backcharge No. _____

Contractor/Vendor _____   Date _____
Description of Need for Backcharge: _____
_____
_____
_____
_____

Applicable Contract/Purchase Order Authority (clauses, sections, etc.)
_____
_____
_____

Backcharge Pricing Method: _____
☐ Lump Sum, or ☐ Estimated Cost: _____
Organization ☐ to perform, ☐ which has performed the backcharged work:
_____

Change Order No. for performance of backcharged work _____
                                                        (if applicable)

*Instructions to Contractor/Vendor:*
The cost of the above-described work ☐ is to be entered and identified as such
                                        on your progress payment estimate
                                        form for the next payment period.
                                     ☐ is shown on the attached invoice.
                                        Please honor by _____ .

Prepared by: _____
               (contract administrator)

Reviewed by: _____
               (construction specialist)

Approved by: _____
               (construction manager)

CONTRACTOR/VENDOR ACCEPTANCE: _____
                                      (contractor/vendor)

Any work performed as a result of this backcharge will not relieve the contractor/
vendor from its guarantee or any of the other terms and conditions of the
referenced contract/purchase order.

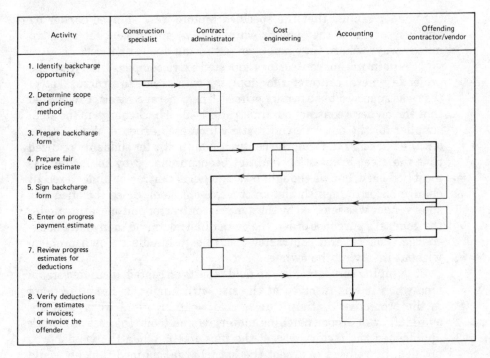

**FIGURE 21.** Backcharging process.

depending on the duration of such work) or crediting contractor invoices when possible.

## REPRESENTATIVE CASES

1.   A vendor of heating, ventilating, and air conditioning equipment was to ship several large fans with associated motors to a heavy industrial construction project. When the shipment arrived, the installation contractor noticed that 32 motors accompanying the fans were of the wrong size and model; they were incompatible with the fans and not in conformance with the purchase order specifications. The construction manager ordered the installation contractor to buy the proper motors from his own sources and begin installation. A cost-plus change order was issued the contractor for his invoiced amount of $16,000 plus $1,600 for profit and overhead. When a backcharge of $17,600 was sent to the fan vendor, he refused acceptance of the charges, claiming that the mistake was not brought to his attention when discovered. After checking,

the vendor replied that the specified motors were shipped earlier by another carrier to the owner's warehouse at an adjacent facility (two miles away), where they had been gathering dust unnoticed.

2. A painting subcontractor requested extra compensation from the project's general contractor for applying prime coats to structural steel that was supposed to be factory primed. The general contractor requested that the owner's contract administrator issue a backcharge to the steel supplier for the painting contractor's increased expense. While filling out the backcharge form, the contract administrator suddenly realized that the steel in question had not been purchased by the owner, as most had been, but was bought by the general contractor under a recent change order to furnish and erect leave-out steel. Since the offending steel vendor was selected by the general contractor and operating under the general's purchase order, the owner had no reason to invoke a backcharge. The general contractor was told to handle its own problem without involving the owner.

3. A piping contractor began field repairs on coated, insulated steam pipe when it was received at the site with numerous handling tears in the protective coating. Due to cold weather, night work, and unavailability of proper patching materials and tools, the cost of rework approached the factory cost of the pipe itself. When the vendor was notified, he refused to accept the backcharge, claimed that he would have gladly shipped new sections if he had been apprised of the problem, and added that the warranty on pipe repaired by the owner was void.

# CHAPTER 15

# CLAIMS

Before beginning a discussion of claims, we must define the term *claim* as it is used in the construction industry. In the context of a construction contract, two principal parties may, and do, present claims against each other:

1. The contractor may claim additional time for performance and/or additional compensation from the owner.
2. The owner may claim relief from the contractor in terms of a reduction in the contract price and/or an acceleration or delay of the contractor's performance.

Of course, there are many other parties, contractual or otherwise, that may present claims to one another or to the owner or contractor during the course of a project. These include subcontractors, consulting firms, the engineering or architectural firms involved, and so forth. Our discussion of claims will center around those most commonly encountered during the performance period—from contractor to owner, or from owner to contractor. The same principles of claims defense or pursuit presented in this context also apply to the majority of other claims situations.

A claim is, then, nothing more than a request, or demand, for cost or time compensation, over and above that which has been granted or contemplated, from one contractual party to the other. Claims may be presented in any number of formats, ranging from an informal or even verbal request to a voluminous, highly structured and heavily documented claims package. A common misconception is that claims are litigious in nature—that is, one party is suing the other for damages in a legal sense. This is generally not the case. Although some claims do fester to the point where resolution between the parties requires a lawsuit or arbitration of some kind, most claims are resolved long before this occurs. The vast majority of claims, be they initiated by owner or contractor, are settled through negotiation, adherence to the terms of contract, or some mutually agreeable adjustment in time and cost of performance among the owner and contractor(s). In this era of litigation and threat of litigation, most reasonable owners and contractors realize that resolution is far more desirable if attained in a nonlitigious manner. Both parties usually suffer when claims progress, or are transformed, into lawsuits.

The goal of those responsible for enforcing and administering construction contracts should be to identify claim situations quickly and resolve them as soon as possible. For owners, the challenge may be put even more simply: Would you rather have a completed project or an enforceable claim in court? Most reasonable owners, and contractors, would choose the former.

## CAUSES OF CLAIMS

Like informal change orders, claims may originate everywhere. There are an infinite number of causes for claims to be found and analyzed in the construction process, but almost all have their basis in the allegation that acts, or omissions, by one party to the contract—or, less often, by third parties, acts of God, or others—caused the claiming party damage of some sort. In the highly complex environment of a construction project, given the time and cost pressure on all parties and realizing the myriad of relationships, responsibilities, duties, and interdependencies, one can easily see why claims are as common a feature of the construction landscape as are concrete and structural steel.

Let us first dispense with the case of an owner claiming against a contractor, for it is by far the less common of the two claims situations mentioned earlier. In general, owners who do file claims against their contractors (and, for that matter, engineering firms or other consultants) do so for one or more of the following reasons:

1. *Defective Work.* Owners not satisfied with the contractor's product may claim damages that include the cost of repair, replacement, or removal of the defective work. In most cases the work does not meet the contractual specifications or is otherwise not fit for its intended purpose. Occasionally, the goods or services in question

do not meet express or implied warranties provided by the contractor or its suppliers.

2. *Delays Caused by the Contractor.*  If a contractor has obligated itself to perform the contract work, as a whole or in part, by a specified time, the owner may claim damages when a delay was caused by the contractor, or in other cases, even if the delay was beyond the control of the contractor. Typical damages claimed by owners in this regard are loss of use of the facility, the ripple effect on other contractors, and the increased cost of other delayed work.

3. *As a Claims Defense.*   Owners faced with claims presented by contractors may retaliate by pressing a counterclaim. This counterclaim usually attacks, or attempts to discredit, the elements of the original contractor claim—by uncovering overlapping or duplicate cost damages, citing contractor contributions to the claim-provoking conditions, or citing changes or claims clauses in the contract that prohibit or modify the contractor's actions in the event of a dispute, for example.

Another type of claim one encounters, though seldom, is that arising from termination or breach of contract. This generally occurs when a contractor fails to complete the work or for some reason leaves the job site. Owners in this situation usually demand to be compensated for the increased cost, over and above that paid to the contractor, for completing the work through other means. Contractors also claim damages when they feel they have been unjustly removed from a project or otherwise prevented from completing their contractual work. Both of these situations arise when the contract, in effect, has been terminated by one party. The focus of our discussion will not be this extreme example, however, for we will concentrate on the more common claims situations—those that arise during the performance period and when continued performance of contractual obligations by both parties is contemplated.

In this regard, most claims found on construction projects are those pressed by contractors against the owner for one reason or the other. Their causes are similar, if not the same, as those for informal changes discussed in Chapter 13. These are summarized below:

1. Late or defective owner-furnished information, generally in the form of drawings or specifications
2. Late or defective owner-furnished material or equipment
3. Changes in regulatory requirements

4. Changed or unknown site conditions
5. The ripple effect of collateral work
6. Restrictions in work method, including delay or acceleration of contractor performance
7. Ambiguous contracts or contract interpretations

In each of these situations, the contractor will claim that something has occurred (or failed to occur) that caused it to incur additional cost or spend additional time beyond that called for in the contract and that could have been reasonably expected at the time of bidding or contract award.

## ELEMENTS OF A CONSTRUCTION CLAIM

If any of the claim-provoking circumstances described earlier occur, a conscientious contractor will immediately notify the owner of the causes and effects of each. When this notification is done by pressing a claim, most contractors will request additional time and/or compensation for: (1) increased costs of performing the changed work; and (2) "impact costs" on unchanged work. In most cases where a bona fide claims situation has occurred, the contractor has suffered some increased costs (be it in terms of time, money, or both) in each of these categories.

It is fairly easy to identify and, for the owner, accept or defend against direct costs of performing changed work. However, the impact costs—cost on unchanged work—are not as easily defined or priced. Let us first deal with the costs of performing changed work. Some of the most commonly cited costs are for:

Increased labor costs (additional or higher-paid workers)
Additional material or equipment required
Additional supervision, administration, and overhead
Increased time required for performance
Ripout and rework
Decreased productivity or efficiency

As for impact costs, all of the preceding may be claimed. The difference is that it is more difficult to determine the nature of the impact and to quantify the increased costs. The question of impact costs can be reduced to a simply put, but difficult to answer, version: What were the increased costs of performing B and C after A was changed? To answer this question, both contractor and owner must determine what it *would* have cost to perform B and C had A *not* changed. This requires a more qualitative analysis and is quite often the most difficult issue confronting those who wish to find a resolution to impact costs.

The best way to illustrate impact costs is through an example. Let us suppose that the owner somehow delays the contractor's performance and causes it to postpone work it planned to perform during the summer months into the winter season. The work itself remains the same, yet the contractor must incur costs associated with winter work over those anticipated for summer performance. These "impact costs" may include:

1.  Cost of protecting the work from the winter weather
2.  Inefficiencies in production caused by workers performing in cold weather, with cumbersome protective equipment, on slippery surfaces, and during shorter periods of daylight
3.  Costs of heaters and fuel for personnel protection and for actual work performance, such as for heating concrete
4.  Increased maintenance costs for equipment
5.  Damage to materials and equipment caused by the weather
6.  Inability to secure a work force
7.  Lost time due to extreme temperature or climatic conditions
8.  Increased housing and transportation costs
9.  Delay due to Christmas holidays
10. Extension of insurance premiums or bond payments
11. Increases in labor, material, equipment, and overhead costs due to inflation and escalation

## THE STRUCTURE OF CONTRACTOR CLAIMS

As mentioned earlier, contractor claims may vary in formality and content. A typical claim, however, generally follows this structure:

1.  A description of the contractual terms and conditions, such as scope of work and pricing structure, covering the segment of work in question
2.  A factual description of events that have (or have not) taken place, usually presented in chronological order and referencing correspondence, change orders, meetings, and so on
3.  The results of the claim-provoking circumstances, usually presented

as a narrative description of the increased effort required of the contractor

4. A cost analysis, which may include a detailed listing of increased costs due to the change or a comparison of actual costs and anticipated costs—the difference comprising the claim amount

One must remember that a claim is different from a contractor's quotation resulting from a notification or change order. In the strictest sense, they may be the same, considering that the contractor is presenting increased cost information to the owner. However, a quotation occurs *before* the work is done, and a claim is generally presented after, or during, performance of the affected work. Once increased compensation or time for performance has been agreed upon, claims should be transformed into change orders.

## CLAIMS ANALYSIS

In order to judge the merits of a claim and to determine what additional compensation, if any, should be allowed the contractor, the owner must thoroughly analyze the claim. This analysis should be done in three parts: (1) a factual analysis (What happened?); (2) a legal or contractual analysis (Is the contractor due relief?); and (3) a cost analysis (How much additional money or time should the contractor be granted?). Factual and legal analyses of claims are facilitated by adherence to the concepts of contractual control described in this volume, that is, proper formation controls, detailed recordkeeping, structured change control, objective progress and payment determinations, and so on. It is surprising, however, how widely cost analyses may vary given the same factual and legal circumstances. It is this area of claims defense— cost analysis—that contains the highest degree of risk and requires the most owner attention.

There are two distinct methods by which contractors seek to quantify the costs of a claim-provoking situation:

1. The total-cost method
2. The incremental-cost method

With the total-cost method, the contractor simply compares the actual cost of performing the work, or work segment, with the anticipated cost (or bid or contract price). The assumption is that *all* increased costs incurred by the contractor are due to the claim-provoking situation.

Needless to say, most owners respond negatively to the total-cost approach. The major problem is that the contractor must prove that the work, as changed, was performed as efficiently as possible. This is difficult to do. Although this approach may provide a useful "upper limit" to the claimed cost, it is generally ineffective in pursuing recovery from the contractor's perspective.

The incremental-cost method is recommended over the total-cost method for several reasons. First of all, it separates increased costs arising from other conditions (contractor inefficiency, bad luck, or factors not pertinent to the claim itself) from those due to the facts of the claim. Secondly, this approach allows costs to be estimated for discrete elements of work under fairly well defined cost parameters. Often with the total-cost philosophy, an unjustified element of increased cost, when included with the claim, obscures or taints the remaining meritable elements, thereby reducing the effectiveness of the claim. And, finally, the incremental method focuses on cause and effect in a one-to-one fashion. With the incremental method, contractors relate each additional cost to each factual cause, for example, "Your direction that soil be hand tamped instead of machine rolled caused us the following increased costs."

## SUGGESTED PROCEDURE FOR HANDLING CLAIMS

As with the change order process, the contract administrator is usually charged with handling claims from their inception through their eventual resolution. And as with several other processes described here, he or she serves more as a coordinator of other project personnel activities than as the authority for the owner.

The first objective of proper contract formation and administration is to eliminate or reduce the incidence of claim-provoking conditions and claims themselves. When a claim is presented, however, the contract administrator should ensure that it is given the proper management visibility within the owner's organization, carefully analyzed and documented, and equitably resolved as soon as possible. Few things are as damaging to the morale of owner and contractor personnel alike as bitter, unresolved claims.

It is particularly important that claim situations be identified and isolated as soon as possible. All conversations, correspondence, supporting documentation, and the like, pertaining to the claim should be obtained and compiled. A separate file should be maintained for each claim and should include copies of those portions of the contract pertaining to the claim, results of factual, legal, and cost analyses, and other documents that may assist in rejecting or resolving the claim.

Each claim received by the owner should be routed to the contract administrator. Upon receipt, the administrator opens the claim file and records the claim in a proposal or claim log for tracking purposes. If additional information is required of the contractor, this should be noted in the log and the contractor notified of receipt of the claim as well as requested to provide the additional information.

The contract administrator then conducts a detailed research effort that includes, among other tasks, interviews with affected owner personnel, review of contract and project files and reports, and compilation of documentation that will be required to analyze the claim. This documentation could include such items as the contract itself, change orders, extra work summary and approval forms, meeting minutes, contractor correspondence, schedules submitted to and by the contractor, project photographs, daily activity or progress reports, time sheets, progress estimates and invoices, and telephone memoranda.

Once the claim file is complete, the contract administrator enlists the assistance of project personnel in analyzing the claim and preparing a response to the contractor. These will include such project participants as the construction manager, design engineers, cost engineers, project schedulers, and the like. Their effort should be structured and coordinated with the goal of an objective analysis and a reasonable response. The contract administrator should never agree with the contractor to increased payments or other relief without proper authority, and it is recommended that the construction manager, or a representative, conduct all negotiations with the contractor. Again, the administrator should maintain his role as advisor to the contractual authority of the owner, not as an enforcer of the contract.

Once a claim has been resolved, a change order should be issued to cover the resolution. In this way, all changes to the contract, be they initiated by the owner (a notification and resulting change order) or by the contractor (a claim and resulting change order) are controlled and documented in the same manner.

For those instances where resolution is impossible without engaging the assistance of higher owner management, the contract administrator should prepare and present the elements of the claim and its analyses to such management and assist them in proceeding as required. Above all, the contract administrator should avoid the tendency to become emotionally involved in the dispute, but should instead accept the challenge to handle each claim on its own merits, to involve and coordinate the efforts of affected or responsible project personnel, and to structure and document the process.

## REPRESENTATIVE CASES

1.  The owner of an industrial complex under construction had recently sent its chief mechanical engineer to a three-day seminar on weld inspection techniques. When he returned, full of enthusiasm for radiographic inspection to detect weld flaws, he revised the specifications for buried process steam piping just after the contractor had commenced work. Since the inspection was to be done by an independent agency hired by the owner, no change order was issued the piping contractor. After several months of work, the piping contractor presented the owner with a claim for $195,000 in additional compensation due to inefficiencies and interference caused by the increased inspection process. Briefly, his claim cited:

The original specs called only for visual weld inspection. The X-ray inspections caused pipe welding to periodically cease in open trenches, as his welders did not want to be exposed to the radiation closeby.

Many more welds were rejected, causing rework and schedule extension into the winter season and resulting in inefficiency in weld production.

Additional welds necessitated by rejects caused him to employ four additional crews. Since qualified welders were scarce in the project location, he paid extra to "import" welders and train or recertify others.

2.  A unit-price contract for a fossil-fired power plant was awarded to an electrical contractor. Drawing changes concerning underground cable ducts and cable runs were issued in a flurry once cable pulling began in the plant. The changes had the overall impact of increasing the length of cable pulled by only 10% over the original estimate, and the contractor was paid by the agreed unit prices for these additions. It submitted a claim, however, for compensation over and above this amount to account for:

Inefficiencies in operation when cable was measured, cut, pulled, and then removed and discarded as revisions to drawings changed cable size and routing

Inefficiencies in cable purchasing and cutting, because the contractor could not plan the usage of cable to maximize usable lengths cut from standard purchased spools

Demobilization, standby, and remobilization of work crews from one area of the plant to others as revised drawings became available

Increased labor cost caused by drawing changes that postponed much of the work into a new labor-contract year, with higher labor rates in effect

3.   A lump-sum contract for installation of the turbine generator for a nuclear power plant was awarded to a mechanical contractor (Contractor A). The equipment was to be delivered by barge two weeks after Contractor A mobilized at the site. A heavy freeze caused ice blockage on the river adjacent to the plant, with a two-month delay in receipt of the owner's turbine generator equipment. To make up for lost time, the project manager ordered the general contractor (Contractor B) to begin installation of 110-inch-diameter circulating water pipe from the turbine building to the cooling tower. When the turbine generator finally arrived, Contractor A could not transfer the heavy components from the barge unloading area to their destination on the turbine pedestal in the turbine building because two 20-foot-deep pipe trenches filled with partially installed circulating water pipe blocked his access. Contractor A filed a claim for increased compensation caused by:

Standby of crews and equipment for the two months of ice delay and additional two months of pipe trench delay

Temporary storage of the turbine generator on site

Acceleration of work once the trenches were covered to compensate for lost time

Loss of profit caused by being unable to use its men and equipment on other jobs

# CHAPTER 16

# BRIEF-FORM CONTRACTING

Not all construction requirements can be seen from the perspective of a home office several months or years before they occur. As a consequence, some flexibility should be given to the owner's site management in order to react to changed or unforeseen events and circumstances requiring the contracting process. Many owners implement brief-form contract documents and procedures for this purpose. Two types of contracting instruments are generally involved: (1) field-issued brief-form contracts; and (2) field change orders to home office–issued contracts.

Without the availability of brief-form contracting instruments, unacceptable contracting methods may be implemented. These include the use of purchase orders, verbal agreements, telephone orders, or the often-abused handshake technique. In order to ensure that all job-site labor is covered by an express, written contract containing protection for the owner, these methods should be avoided in all cases. But when the alternative is a lengthy formation process or unnecessarily complicated documents, unacceptable contracting methods often occur. The need exists for an alternative method of contracting that is rapidly achieved and easily administered and at the same time provides formation and administration controls essential to the owner.

## BRIEF-FORM CONTRACTS

The major objection to home office–issued contracts is that their formation takes a long time. Several months are often required to qualify bidders, create the RFP, prepare bids, and receive and evaluate proposals before an award can be made. In addition, "long-form" contract documents often contain many clauses that are unnecessary for short-term, low-cost scopes of work. Long-form contracts should therefore be reserved for all foreseeable work and for scopes of work that are medium to large in amount, risk, and cost.

Brief-form contracts are designed to apply to other conditions. They are used for short-duration, low-cost work that cannot be planned and priced well in advance. Unforeseen events such as emergencies and planning errors fall into this category. Where the typical long-form contract is represented by several hundred pages of documents, a brief-form version may be written on a dozen or less. Bidder qualification and other formation processes are performed in an abbreviated fashion, and a contract agreement may be formed in a few days or less, rather than over a period of several months. Common applications for brief-form contracts may include:

1. *Emergency Work.* Repairs caused by fires, storm damage, or safety measures fall into this category.
2. *Housekeeping Services.* These include general cleanup and maintenance services performed during the construction period. Examples are snow removal, trash collection and disposal, janitorial services, guard services, and sanitary services.
3. *As Relief for Large Contractors.* Should an existing contractor be unable or unwilling to perform additional work, smaller, local contractors may be hired to perform the work.
4. *As Competition for Change Orders to Existing Contractors.* In some cases cost quotations from existing contractors to perform additional work may appear to be noncompetitive or excessively high. If the work may be performed by other, locally available contractors, they should also be asked to quote. This provides some means of protection from an existing contractor that feels it holds the owner captive by its presence at the job site.
5. *When Existing Contractors Have Been Performing Unsatisfactorily.* Should additional work be required and the owner is dissatisfied with the available sources at the site, brief-form contracts may be used to hire other contractors capable of better performance.
6. *To Perform Backcharge Work.* At times there is no site contractor available or capable of performing backcharged work.

Although their use must be carefully controlled, brief-form contracts have many significant advantages. They can be rapidly formed. The cost of formation is often insignificant when compared to the cost of forming long-form contracts. They are ideal for small scopes of work, where the formation of complex contractual documents exceeds or is out of proportion to the work to be performed. They are excellent when used for the rental of construction equipment. They can be used to relieve large contractors from the requirements of standing down or remobilization to complete minor sections of work.


## PROCEDURE FOR BRIEF-FORM CONTRACTS

Brief-form contracts should be formed and administered with the essential elements of control required for long-form versions. Competitive bidding by qualified contractors should be encouraged, though the degree of qualification and the number of available bidders are generally reduced. Standard or reference contract documents should also be prepared, each containing essential clauses found in their long-form counterparts. And administration documentation should be maintained throughout the performance period. In order to control the formation and administration of brief-form contracts, the following suggestions should be considered:

1. To the extent practical, brief-form documents should closely parallel the content and format of long-form contracts.
2. Competitive bidding should be encouraged if possible.
3. Existing contracts, or those planned for the future, should be reviewed before the need for brief-form contracting is established.
4. The alternative of adding work to an existing contract (change order) should not be overlooked when planning a brief-form contract.
5. Close coordination between long-form contracts, typically formed in the home office, and brief-form contracts, typically formed at the site location, should be maintained to avoid duplication of effort, redundant scopes of work, or loss of scope entirely.
6. Administration of brief-form contracts should not be conducted differently than that of other contracts.
7. Limitations on the use of brief-form contracts should be established and enforced. When brief-form contracting is allowed, most owners stipulate certain cost, time, or scope limitations on its use. For example, any contract priced at over $50,000 may require the use of long-form documents and the standard formation processes described in Part 2. A common tendency, however, is for these "thresholds" to be exceeded—for brief-form contracting to be used for larger cost items. When this occurs, owners should make sure that the thresholds bear a resemblance to actual project conditions and needs. If they are unreasonably low, they should be raised. If

they are sufficiently high, they should be enforced. And actual contracts should be reviewed to ensure that a series of brief-form contracts is not being used to achieve results that could have best been accomplished by their consolidation into one larger, long-form version.

## PROTECTION VERSUS BREVITY

For all the benefits of brief-form contracts, there is the risk that protection is lost by virtue of their condensed form. Needless to say, if any contract clauses were unnecessary, they would also be eliminated from long-form documents. So why are brief-form documents "more brief" than long-form versions? The answer lies with the fact that long-form documents are used for special conditions not envisioned by brief-form versions. Some examples of provisions required by long-form contracts that may be totally unnecessary in shorter versions are:

1. *Progress Payments.* Because brief-form documents are used for short-duration work, the provisions covering the measurement and payment for progress may not be required; final payment may be all that is involved.
2. *Changed Work.* Since the nature of brief-form contracting is generally to cover changed work, changes to brief-form contracts may not be required. Should additional work be necessary, it may be covered by a change order to another contractor or through the use of another brief-form contract.
3. *Contractor Submittals.* These may not be required.
4. *Equipment Warranties, Titles, Spare Parts, Tools, and so on.* Most brief-form contracts are service related, without the need for complex clauses to protect the owner when equipment and materials are also bought.
5. *Closeout Documentation.* Most likely, the brief-form contractor is not supplying as-built drawings, releases of lien, or other documents upon completion of the work.
6. *Technical Specifications and Drawings.* Brief-form contracts are generally reserved for low-technology work, and there may be no need for technical specs or drawings to define the scope of work. Where used, the specifications may entail just a few sentences or paragraphs.
7. *Required Contractor Controls.* Sophisticated cost and schedule

control systems, planning of work, progress reporting, and cost records may not be required.

8. *Retention.* Because of duration and scope of work, this may not be employed.

These are just some of the areas in which brief-form contracts may be streamlined. Significant danger lies in using such contracts when the cost, scope of work, or contract duration exceeds that envisioned by the brief-form documents. Because of their coverage limitations, brief-form contracts should never be used when they do not contain provisions for anticipated or potential needs during the performance period.

## BRIEF-FORM CHANGE ORDERS

Owners often employ two distinct versions of change order documents for the same reasons that they use two versions of contract documents. Brief-form change orders may be identified and employed for field-originated changes, emergency work, or minor alterations to the contractor's scope. Long-form change orders are often restricted to changes originating in the home office, such as design revisions, for changes that are planned well in advance of incorporation, and for changes requiring significant time to consider, evaluate, and approve. Brief-form change orders are then used for the remaining change situations—those that are unplanned, result from site-specific conditions (bad weather, interferences, and so on), and cover immediate needs caused by unforeseen events.

Just as with brief-form contracts, brief-form change orders are designed to prevent unacceptable contracting practices that may be used to respond to unforeseen and pressing events. They shold be identified, evaluated, and approved with the same degree of control as their home office–issued counterparts. Particular attention should be given to the possibility of duplication between home office change orders, field office change orders, home office contracts and field office (brief-form) contracts.

## BRIEF-FORM CONTRACTING CAUTIONS

For all of their advantages, both brief-form contracts and change orders contain special dangers and are often misused for the sake of expediency at the expense of control. They should be carefully designed and limited to specific applications, and their use should be continually monitored.

Because of their special nature and the environments under which they are generally used, certain additional cautions apply to owners who employ them:

1.  Because brief-form contracts are often used with smaller, local contractors, pay particular attention to insurance coverage, union affiliations, if applicable, and bonding capacity, where required.
2.  Establish procedures to eliminate possible duplication of effort and payment between short- and long-form applications.
3.  Review short-form documents prior to each application to ensure that critical contractual protection is not lost by the abbreviated nature of the documents.
4.  Avoid the tendency to price brief-form contracts or change orders in a cost-reimbursable manner when other fixed-price methods may be used.
5.  Monitor adherence to time, cost, or scope limitations.

## REPRESENTATIVE CASES

1.  On a project to build an addition to an automobile assembly plant, a fire broke out over a weekend and gutted the facility's existing warehouse. Since the warehouse was needed for storage of equipment and material to be used in constructing the addition, it was necessary to demolish the gutted portions and make temporary repairs as soon as possible. A field contract was issued to a local general contractor to perform this work.

2.  Due to expansion of site office staff, the construction manager for a pulp mill project decided that additional temporary parking areas adjacent to the office trailers were needed. Upon checking the local Yellow Pages, the contract administrator found a grading and paving contractor willing to do the work for $12,000. The response to a notification to the site civil contractor was a quote of $26,000 if this work was to be made a change order. The local paving and grading contractor was awarded the work by a field contract for $12,000, and no change orders were issued to the site civil contractor.

3.  Sheet pile driving at a process plant's intake water pumping station was supposed to take place during the dry season. Heavy rains and the failure of a small upstream dam flooded the area after the piles were driven but before final inspection of pile splices could be made. Underwater divers were required to make this inspection, and the

company providing these specialty services was engaged through a field contract.

4.   A short oil pipeline project required the crossing of an unguarded railroad line by material delivery trucks. For safety reasons, the owner decided to man the crossing with a flagman during working hours. A guard service was hired to provide flagmen under a field contract.

5.   The owner of a nuclear power plant under construction decided to hire certain construction cranes for use in the area of the reactor building rather than to require several contractors to provide their own. The owner scheduled their use and prorated their cost among the using contractors. The fully operated cranes were rented under a field-issued equipment rental contract.

6.   The painting contractor for a large industrial plant project had completed all of his painting except for a small masonry auxilliary building that had been delayed due to a shortage of concrete blocks in the area. Rather than keeping the painting contractor on standby at $500 per day, the owner closed out his contract with a credit of $6,000 for not painting the building in question. Four months later, the building was constructed, and a local contractor was awarded an $8,500 field contract to paint it.

# CHAPTER 17

# CONTRACT CLOSEOUT

Construction contracts should be formally terminated once performance has been completed and final payment has been made by the owner. Because final payment and the release of remaining retention constitute the owner's release of its most vital compliance control (the payment of money), a structured process leading to this release is recommended. This process, called contract closeout, involves final acceptance of the contractor's work, receipt of acceptable documentation required by the contractual terms, and an evaluation of the contractor's performance. Though it is performed with the assistance of many people, the contract administrator should coordinate and control contract closeout. This chapter describes the closeout process, discusses the responsibilities of those typically involved, and presents a recommended procedure for its conduct.

## DOCUMENTATION INVOLVED

Depending upon the contractor's scope of work, the owner's objectives, and the specifics of the contract in question, closeout generally involves the receipt of documentation from the contractor. In order to ensure that such documentation is actually submitted and is acceptable, final payment and release of retention (if used) should be withheld until it is received and approved. Commonly required documentation is described in the following paragraphs.

1. *Releases of Liens from the Contractor and Its Suppliers and Sub-contractors.* These releases are formal documents signed by those who have performed work or provided materials or services in conjunction with the contract. They are, in effect, promises that no liens have been

filed or will be filed. Their format and content should follow any applicable legal requirements. Some owners insert sample release forms in the contract documents at the time of award. These releases constitute strong protection for the owner for two reasons: (1) they may be enforceable in court when a subsequent lien is filed; and (2) when made a prerequisite to final payment, they force the contractor (and its suppliers and subcontractors) to examine their books in order to detect any lack of payment or failure to receive payment. Should either have occurred, it is much better to present these problems at the time of closeout than afterward. They can then, it is hoped, be rectified before conditions leading to a lien situation arise and interfere or prevent the owner's occupation or use of the completed facility.

2. *Titles to Major Equipment Incorporated in the Facility.* Ownership of certain pieces of equipment and machinery may need to be transferred to the owner through clear titles. These should be received and verified before final payment is made.

3. *Warranty Documentation.* Operating equipment is generally sold with a specific warranty, usually in effect for one year. The warranty documentation for this equipment should pass to the owner in accordance with customary practices or specific contractual terms.

4. *As-Built Drawings.* Most major facilities are not constructed in exact accordance with the drawings issued prior to or during the construction effort. This is not to suggest that drawings are not followed, for they should be. There are many cases, however, where the drawings do not show exact locations, positions, relationships, and sizes. Instead, they give general locations, suggested positions, and general dimensions—allowing the constructor tolerances and flexibility in installation or erection. In other cases, changes to the work, for the benefit of the owner or to simplify construction itself, are made as the construction process is carried out. When these changes depart from the drawings, actual as-constructed positions or locations should be shown on a separate set of drawings called "as-builts." A common example of as-built conditions that differ from design or suggested conditions involves small diameter piping that is routed according to field conditions at the time of its installation. Most design drawings leave this routing to the contractor. Naturally, the owner should check the routing to make sure that there are no interferences with other objects and that the piping is supported soundly, achieves its purpose, and is installed in a workmanlike manner. But the details of its installation are lost unless described by the as-builts. Other examples include minor adjustments for field fit up of equipment, conduit, cable tray, ductwork, and the like. As-built drawings are important. They often represent the only

accurate record of the constructed facility and are often needed for operations, maintenance, and repair throughout the facility's life. Should renovations or retrofits be needed years later, the as-builts allow these to be designed and constructed with full knowledge of what is "out there." Their importance is heightened when one considers that much of the completed work may be buried underground, enclosed, or otherwise inaccessible. Millions of dollars have been spent locating and measuring such items when representative as-builts were not available. Three specific recommendations are given for dealing with as-built drawings. They should be thoroughly checked and verified before approval. Should the cost of their preparation be included in the contract price, make sure that the owner does not incur this cost a second or third time. This can happen when the engineering firm or others prepares as-builts instead of (or concurrently with) the contractor. In addition, approved as-builts should be secured and protected once received, to be easily located and understood years later.

5. *Inspection and Acceptance Records.*   These include test reports, calibration data, and inspection results. Although initially required to ensure that the work was performed correctly and that the finished products are acceptable, they also provide valuable information that may be needed at a future date.

6. *Operating and Maintenance Manuals.*   The need to operate and maintain the finished facility is often overlooked in the rush to complete construction and closeout the contract(s). Owners should receive appropriate documents needed for both purposes. Training material relating to the actual operation of completed systems, processes, or equipment should also be received where needed.

## OTHER TYPICAL DELIVERIES

In addition to closeout documentation, other items may also be required before the contractor is released of its contractual obligation. Examples include:

1. *Spare Parts.*   These are especially important for specialized equipment and equipment that is custom made for the project. Critical equipment and equipment prone to failure should be furnished with some spare parts; the owner should identify these and make their receipt part of the contract requirements. This is especially true for equipment having long lead times to replace or repair and equipment supplied by only a few sources.
2. *Special Tools.*   Many constructed or installed items require special

tools not normally maintained by the owner. These should be identified, required by the contract, and received prior to final payment.

3. *Consumable Supplies.* This category might include special fuels, lubricants, gaskets, seals, valve packing, electrical parts (light bulbs), cleaning solvents, and liquid or gaseous charges.

Needless to say, it is difficult or impossible to obtain spare parts, special tools, or consumable supplies unless they are specified by the original contract documents. They should be considered not only when the contractor is about to depart but also as part of the contract formation process. If continuing service agreements are made part of the construction contract, the prices of both services and materials should be covered as well.

## SUGGESTED CLOSEOUT PROCEDURE

Owner control over the construction process typically declines toward the end of the project. This is generally because of the schedule pressure commonly encountered, the reduction of supervisory personnel as the work winds down, and a commonly felt urgency to finish the work and move on to newer challenges. Unfortunately, however, this is the time when control is perhaps needed most. In order to manage the closeout process, owners should design and implement structured procedures well in advance of this period. Such procedures, when supported by a well-designed and properly used closeout checklist, help ensure contractual compliance during the final stages of the performance period.

Exhibit 20 depicts a general closeout checklist that is designed to ensure that all required items have been received, all work has been

---

**EXHIBIT 20.   Contract Closeout Checklist**

---

Contract Closeout Checklist

Owner

Project

Contract No. _____
Contract Work _____
Contractor _____

*(continued on p. 234)*

**EXHIBIT 20**  *(continued)*

| Check Item | Yes | No | N/A | Initial |
|---|---|---|---|---|
| 1. Engineer final inspection | | | | |
| 2. Corrections complete | | | | |
| 3. As-built drawings received | | | | |
| 4. As-built drawings satisfactory | | | | |
| 5. Operating manuals received | | | | |
| 6. Warranty statements received | | | | |
| 7. QA/QC documentation complete | | | | |
| 8. Backcharges satisfied | | | | |
| 9. Release from contractor | | | | |
| 10. Final invoice approved | | | | |

| Route: | Signature | Date |
|---|---|---|
| 1. Contract administrator | | |
| 2. Cost engineer | | |
| 3. Resident engineer | | |
| 4. Construction accounting | | |
| 5. Construction manager | | |

accepted, and no commercial transactions are incomplete. It should be completed by owner personnel and representatives responsible for verifying specific elements of the contractor's performance—for example, the resident engineer checks the as-builts, the contract administrator checks such items as backcharges and unresolved change orders, and construction accounting reviews the final invoice. The completed checklist should be a prerequisite to final payment along with the final progress payment estimate form submitted by the contractor.

## POSTPERFORMANCE EVALUATION

As part of the closeout process, a thorough and objective evaluation of the contractor's performance should be made. This evaluation should be made by those who have dealt with the contractor in many different areas, including the construction specialist, contract administrator, and construction manager. Completed evaluation forms (see Exhibit 21) are then returned to the owner's bidder qualification files for use when reviewing the contractor for future work. Owners should ensure that these forms reflect the entire scope of a contractor's performance, that they assist in making an objective evaluation, and that any problem areas are thoroughly explained. Completed evaluations should not be transmitted to the contractor but should be used only for owner information.

It is also suggested that field monitoring and administration personnel indicate improvements they recommend for future contracts of the same type. For example, if the contract in question lacked commercial controls, contained obscure or conflicting requirements, or was priced in a manner that made determination of progress difficult, these should be noted. In this way, the contractor evaluation becomes an indicator of not only the contractor's performance but also of the contract itself. This will, it is hoped, allow improvements in future contracting efforts.

---

**EXHIBIT 21.   Contractor Evaluation**

---

CONTRACTOR EVALUATION

Date _____

Contractor _____        Project _____

Address _____        Contract No. _____

Title _____

Contractor's Personnel:                        Report Period _____ To _____

Project Manager _____        Type of Contract _____

Project Engineer _____        Contract Amount: Original $ _____

Field Superintendent _____        Change Orders: No. _____ $ ____

Scope of Work Performed _____        Final Contract Amount: $ _____

Instructions: Contract Administrator in cooperation with the Construction Specialist prepares the evaluation for review and approval by the Construction Manager.

| Item | Satisfactory | Unsatisfactory |
|---|---|---|
| 1. Supervision | _____ | _____ |
| 2. Planning | _____ | _____ |
| 3. Support from home office | _____ | _____ |
| 4. Drawings | _____ | _____ |
| 5. Material delivery | _____ | _____ |
| 6. Construction equipment | _____ | _____ |
| 7. Labor | _____ | _____ |
| 8. Progress payment preparation | _____ | _____ |
| 9. Administration of extra work | _____ | _____ |
| 10. Cooperation | _____ | _____ |
| Composite rating | _____ | _____ |

*(continued on p. 236)*

**EXHIBIT 21** *(continued)*

Explain unsatisfactory items (use item number):

Was an undue amount of time spent administering this contract?

What improvements could have been made in this contract?

Would you recommend awarding future contracts to this firm?
Additional comments:

Prepared By _____      Approved _____
                    Contract Administrator                                    Construction Manager

                    _____
                        Construction Specialist

## TERMINATION NOTIFICATION

Once the work has been performed and accepted, the closeout checklist
completed, and an evaluation made, the owner should consider the
need for a formal letter of termination. This letter informs the contractor
that the contractual association is officially ended; it may be sent to
others who have participated in the relationship as well. Formal notification is recommended for a number of reasons. First of all, it officially
notifies the contractor that charges under the contract will no longer
be accepted by the owner—and that no additional work should therefore
be performed. Copies should be sent to owner accounting or accounts
payable groups for this reason. If special billing or account numbers
have been used by the contractor or owner personnel, these should also
be officially terminated. This prevents unwarranted charging of time
or other costs to accounts that should not be used. If unresolved claims
or change orders still exist, the termination letter should identify these
and describe the owner's plans for their resolution. The amount of and
reasons for any remaining retention or withholding of payments should
also be identified.

Completion of contract closeout signals the end of the contractual
relationship between the owner and the contractor and, as such, the

final step in the commercial administration of the contract. When performed under controlled conditions and formally documented, closeout ensures that both parties have performed according to the agreement and that both have received the benefits of the relationship. The owner benefits because it receives the necessary documentation and deliverables required to complete the construction process and operate and maintain the constructed facility. The owner also benefits by receiving valuable information regarding both the contractor's performance and the performance of the contract documents used. Both should improve future contracting efforts. The contractor benefits by rapid resolution of remaining items that may impede receipt of final payment. Third parties are notified of the termination in order to prevent unwarranted charges or the performance of work outside of the contractual agreement. And the relationship that began under structured and formal conditions (the signing of a contract) is dissolved with the same degree of control for the benefit of all concerned.

## REPRESENTATIVE CASES

1. Before approving the final invoice and releasing retention for a large piping and mechanical contract, the field accountant conducted several routine checks. He called the field engineer and was told that a roll of as-built drawings had been received. The plant operations department told him a shipment of spare parts and special tools for the installed equipment had been received at the plant warehouse, and the contract administrator informed him that a release of liens form from the contractor was "in the mail." The field accountant vouched the invoice, and in two days the contractor picked up his check and departed. One week later, the project manager was on the phone to the accountant demanding to know why final payment was made. It seems the as-built drawings were no more than original contract drawings, each stamped "As-built" by the contractor. No modifications due to change orders, field routing of small bore pipe, or changes in underground pipe routing and elevations due to interferences were shown on the drawings. In addition, liens had just been filed in the county courthouse by an unpaid pipe supplier and an excavation subcontractor. And finally, the boxes of spare parts and special tools received by the warehouse were opened. They contained a crescent wrench, a can of 3-in-1 oil, and a book of matches advertising "Learn to be a mechanic for fun and profit."

2. Mr. X, civil superintendent for the construction of a chemical processing plant, believed that he should be the one to file a contractor

evaluation report on the site civil contractor—after all, X had been in charge of the contract. X had no problem with Contractor A—they had worked well together over the past two years. Contractor A had remembered X's birthday with a case of premium bourbon, and each Christmas and Thanksgiving his automobile trunk was certain to contain a huge turkey courtesy of Contractor A. Besides, X and Contractor A's superintendent went way back together, and X had been given an open job offer by Contractor A any time he tired of his present position.

X gave Contractor A outstanding ratings for its performance and sent copies of the evaluation to each division of the owner's operations. Had he checked with others on the project staff, he would have learned that:

Contractor A was dependably belligerent and demanding when negotiating change orders.

Contractor A, as a habit, was five months late in paying subcontractors. These subs constantly complained to the owner and threatened to cease work many times.

Contractor A's invoices were never in agreement with progress estimates and usually contained clerical errors, overbillings, and duplicate markups for expenses.

Contractor A refused to remove two foremen caught selling narcotics on the job site.

Contractor A's shop drawings were of poor quality and consistently late, causing constant rescheduling of other contractors' work.

Contractor A's general liability insurance had been canceled near the end of the job due to repeated accidents on all its projects, as well as countless violations of safety regulations.

Contractor A's construction equipment was continually under repair or not operating due to lack of preventative maintenance and qualified operators.

# PART 4

# CONTRACT MONITORING

Without regard for the organizational strategy, pricing methods employed, or even the methodology surrounding the performance of contract formation and administration for a project, there exists a need for management to monitor these processes. Proper contract planning, the formation of contracts containing compliance controls, and effective contract administration can be implemented at the very beginning of a project—but by which means does company management assure itself of continuing control, identify problems that may jeopardize project success, or take corrective action when those problems are identified?

*Contract monitoring* is a term that applies to those systems and activities designed to answer this question. It entails a formal program of contract auditing and reporting, along with a carefully established organizational approach to the performance of contract formation and administration functions. This part will address those techniques that, along with those already presented, allow management to maintain visibility over contracting efforts. It contains discussion of:

*Functional Contract Support.* As opposed to more general project responsibility assignment issues, this chapter analyzes the role of the contract support function in the project environment and addresses typical contract organizations, their benefits, and disadvantages.

**239**

*Contract Reporting.* Several reporting requirements will be identified, and suggested reporting formats, frequency, and information-gathering challenges will be presented.

*Contract Auditing.* This chapter discusses the role of internal and independent auditors and presents distinctions between the need for and conduct of operational as well as cost audits.

# CHAPTER 18

# FUNCTIONAL
# CONTRACT SUPPORT

After reviewing the major functions grouped under the general topics of contract formation and administration, we now need an understanding of the approaches owners take to perform these functions. As mentioned earlier, the structure of an owner's project organization will vary widely depending on the contractual strategy employed, owner's degree of involvement, and individual preferences, among other factors. This chapter examines the benefits and disadvantages of the most commonly used organizational approaches as they pertain to contract formation and administration.

## PROJECT ORGANIZATIONAL SETTINGS

There are an infinite number of ways a large construction project may be organized, and the owner's involvement can vary from a minor one, as with a design–build, turnkey project, to exclusive control of the project, such as one built by force account. In addition, the owner may place agents, representatives, or contracted management and staffing personnel in its project organization in place of its own personnel.

It is difficult, if not impossible, to describe in any detail the functions of contract administration for each of the thousands of organizations an owner may create to monitor and manage a project being constructed by one or scores of construction contractors. In order to treat this subject with as much detail and consideration as it demands, as well as to avoid parochial comment or viewpoints, we will assume a generic or functional project organization. Whether the positions are filled exclusively by owner personnel or by contracted management or staff to some degree (which is usually the case) has been irrelevant to our discussion.

Every owner will appoint a full- or part-time "project manager" representing it for all project considerations. Under him or her, we usually see one owner representative over the construction phase of the project, the construction manager. On most large projects, the construction manager will establish a temporary office and locate his staff of support personnel at the construction site location, the site office. Preceding and concurrent with construction, design engineering and procurement, including contract formation, is usually accomplished in the permanent offices of the owner, A–E, and/or contracted project management organization—the home office.

The field staff, at the site office, consists of personnel responsible for several activities and functions in the interest of the owner. Typical generic positions and their duties or areas of action are:

*Construction Specialists.* They are responsible for monitoring the technical performance of contractors at the site. Called area engineers, area superintendents, discipline engineers, or superintendents, each is assigned a group of contractors, a construction discipline (civil, electrical, mechanical, and so on), or physical work areas (liquefaction plant, turbine building, yard structures, and so on).

*Resident Engineers.* These field engineering personnel are usually divided by engineering discipline. They may perform engineering–design services needed as construction proceeds and act as liaison between the site office and engineering staffs in the home office.

*Contract Administrator(s).* This person performs contract administration functions at the site office.

*Construction Accountant.* This position is responsible for payroll, invoice receipt and payment, and collection of actual cost data.

*Construction Cost Engineers.* They prepare cost reports and cash-flow reports and perform estimates.

A typical organization chart for the project organization, showing both home and site offices, is shown in Figure 22. Of course, many other positions can be and are represented in some field office organizations. These may include inspectors, surveyors, labor relations specialists, safety engineers, office administration, material control specialists, guards, warehousemen, document control clerks, and those responsible for scheduling, quality assurance, field purchasing, and plant operations,

**FIGURE 22.**   Sample project organization.

to name only a few. Only those typical positions with direct interface
to contract administration are identified in Figure 22. On many projects,
these functions can be grouped so that one person is responsible for
several. In this example organization, site contractors communicate
with the construction manager as the chief owner's representative. All
instructions, orders, notices, and so on, from the owner to contractors
and vice versa are issued by or to the construction manager.

## TYPICAL ORGANIZATIONAL APPROACHES FOR CONTRACT FORMATION AND ADMINISTRATION FUNCTIONS

Few companies allow a centralized group, such as a contract administration department, to exercise contract controls exclusively or execute
contracting processes. There is usually a need for two types of functions:
(1) localized project or performance management (contract "users");
and (2) a centralized contract support group serving the needs of these
contract users.

Since the functions of contract formation and administration have been somewhat unidentified in the past, many companies are wrestling with the identity of such functions, the quality and expertise required of personnel to perform these functions, and the organizational setting in which they will operate. No organization is the same, but most sophisticated companies are choosing an approach that provides for contract support as a staff function, serving the needs of varied contract users, or projects. In this manner, contract specialists—the contract administrators—reside with a centralized staff organization yet serve, or are assigned to, specific user organizations. Common approaches using this, or other rationale, are described in the following subsections.

## THE PURCHASING APPROACH

With this approach an owner essentially views contract formation and administration as elements of a larger procurement, or purchasing, function. It assigns responsibility for contract formation and administration to an existing purchasing department. The purchasing department may delegate the responsibility to a designated group of buyers and purchasing agents or assign all purchasing duties, including contract work, on a project basis. Although this approach may on its surface appear logical—contract management may be seen as only a particular type of purchasing—the complexities and added precision required for construction contract "purchasing" may place it beyond the understanding and ability of some purchasing organizations. Specific problems with this approach are:

1. Purchasing personnel may lack engineering and construction experience and training. This lack of understanding of the environment, processes, and practices of construction can be a major handicap.
2. Purchasing procedures and documents, although suitable for ordering supplies and equipment, are normally not suitable for construction contract management.
3. Purchasing personnel administering contracts from a home office are physically removed and therefore may be unresponsive to the day-to-day demands of the construction project.

## THE CONSTRUCTION ENGINEER APPROACH

This is a common approach in which the functions of contract administration are not distinguished from those of performance monitoring. As such, the construction specialist (engineer, superintendent, or area engineer) is responsible for the administration of the contracts under his or her jurisdiction (area or discipline) as well as coordination and oversight of the schedule and technical performance of the affected contractors. This may potentially represent a classic conflict of interest, as described earlier, and should be avoided on all but the smallest of projects. Specific problems with the construction engineer approach are:

1. Conflicts of interest between the functions of performance monitoring and contract administration are likely. Just as church and state, these two functions should remain separate.
2. Construction specialists may lack exposure to the contract formation process and understanding of the contents and intent of legal or commercial aspects of construction contracts.
3. Most construction specialists are adverse to the requirements of paperwork associated with contract administration.
4. Use of construction specialists to perform contract administration is a misapplication of skills. It is better to allow the specialist to do what he or she is most capable of—technical monitoring.
5. Contract administration performed by several practitioners soon becomes inconsistent, disjointed, and fragmented.

## THE PROJECT OR CONSTRUCTION MANAGER APPROACH

The project or construction manager should be above the details of contract administration. He or she is the ultimate enforcer of contract requirements and should use the contract administrator in a support staff function, much the same way he relies on other specialists in areas such as engineering, scheduling, and accounting.

## THE COMMITTEE APPROACH

No list would be complete without mentioning the committee approach. Here, contract management is seen not as a distinct specialty but merely as an adjunct set of duties to be performed by several people in addition to their mainline functions. Specific problem areas here involve the following:

1. There is a lack of consistency and uniformity among contract documents, processes, and procedures.

2. There is confusion, misunderstanding, and error in contract formation and interpretation.
3. There is an inability to determine responsibility (or culpability) for contract formation and administration duties.
4. Committees are generally unresponsive to pressing needs on a timely basis. In times of crisis, rapid and decisive action is difficult to obtain.

## SUGGESTED APPROACH

As can be sensed from the preceding chapters, contract formation and administration represent distinct functions essential to the completion of a construction project. As such, they should be performed by specialists responsible for specific duties and equipped with the tools, training, and authority required. On large projects, a full-time contract administrator(s) should report directly to the project or construction manager and be chartered to perform the formation and administration duties described in this volume. He or she should be supported by a functional group (contracts department) that provides technical and administrative support to each construction project. Owners who cannot justify this commitment over a period of time should consider contract assistance from A–E or contract management consulting firms on a project basis.

# CHAPTER 19

# CONTRACT REPORTING

Over the course of a construction project, owners need key information to determine the status, performance, and expected results of critical project activities. A multitude of reports have been designed and implemented for this purpose. Among these are procurement reports, cash-flow projections, schedule progress reports, cost forecasts, labor usage reports, productivity reports, and the like. These fall into the general category of "project reports" and are used to give owners or their representatives both summary and detailed information.

Contract reporting deals exclusively with parameters that indicate the commercial success, or failure, of the contracting process. As such, contract reports should be considered as a specialized category of reports within an owner's overall project information and reporting system. The relationship between contract reports and other project reporting efforts is sometimes difficult to establish. Many times contract-specific reporting systems overlap with other project reports and/or conflict with information presented by them. In the design and use of contract reports, a major challenge is to identify key performance indicators over the contracting process, present them to management in a way that allows isolation of problem areas for corrective action, and to ensure that this information is reconciled or integrated with other data used for the project.

## PURPOSE OF CONTRACT REPORTING

For most owners, a large portion of the cost of a new facility is committed and paid for contractually. This means that direct monitoring of contractual cost data is essential for control not only of contract costs in themselves but also, by controlling contract cost, of the cost of the entire project. Contractual cost information, therefore, is a major element of contract reporting. Owners rely on the schedule performance of contractors in order to ensure that the facility is completed on time. Project schedule performance is therefore directly related to the timely performance of contractors and their subcontractors. Because technical performance—the quality of a contractor's work—is the subject of other technical monitoring and reporting efforts, contract reporting should focus on the two remaining elements of risk: cost and schedule.

Reporting of information always takes a great deal of effort and results in some added cost. Any contract reporting system envisioned by owners should be designed and implemented so that the cost of gaining contractual information does not outweigh the benefit of having such information. In order for this to occur, the reporting objectives described in the following paragraphs must be met.

1. *Status and performance information should be presented in a format that allows analysis.* Nothing is more frustrating and less valuable than information that does not facilitate problem analysis and subsequent corrective action. Well-designed contract reports should contain easily comparable data regarding planned (or budgeted), actual, and forecast performance. In the case of costs, contract reports should show *estimated contract prices, committed prices* (that is, the contract price once it is awarded), *actual contract prices* (paid to date, and as amended by change orders, backcharges, credits to the owner, and so on), and *estimates of final price* upon completion. These data allow the owner to determine what was expected, what was achieved through formation, what has happened so far, and what is expected to happen in the future. Significant discrepancies among these should point out areas for further investigation, for they indicate that critical problems have occurred or are about to occur. The most common problem hampering this type of analysis is caused by differences in the structure or format of cost data. Every effort should be made to ensure that contract prices are arranged in the same format throughout the project life, from the structure of prices requested in the RFP to the "line items" on progress payments and the entries on contractor invoices. Otherwise, meaningful comparisons between planned, actual, and forecast amounts are difficult or impossible to obtain. Because contract reports typically use the same cost data as other project cost reports, or are even used as input to these reports, they should be structured to allow relatively easy comparison between the two.

2. *Reports should allow monitoring of contractor progress.* Many progress reports depict activities on the part of the owner yet do not

give a complete picture of contractor progress. This is particularly true with regard to schedule and cost information. For example, a report showing that of a total contract price of $1 million the owner has paid $500,000 to date indicates the owner has spent one-half of the contract amount. This tells nothing of the contractor's progress. It could be that the contractor has completed one-third of required work, even though it has been paid for half. The use of discrete, objective progress payment techniques helps to solve this type of problem, but allowances should also be made for the impacts of escalation, credits to the owner, back-charges, change orders, and work that is partially completed yet not paid for—not to mention the time lag between a contractor's achievement of progress, acceptance by the owner, and eventual payment for that progress (the invoicing cycle). To be meaningful, therefore, comparisons of planned with actual performance should be precise and well planned and made on a current basis. Instead of comparing current payments with total contract price, then, current performance should be compared with planned performance for the same period.

3. *Contract reports should focus on owner performance as well.* As mentioned earlier, compliance with contractual terms is required of both parties. The most critical, and often overlooked, element of owner performance relates to the timing of owner activities. For this reason, reports should compare scheduled owner performance, such as review of design drawings, review and approval of contractor submittals, and completion of acceptance testing, with actual owner performance. Significant delays between planned, or committed, and actual owner performance could lead to: (1) delay in contractor performance; (2) increased cost of performance through constructive change orders; or (3) contractor claims. Not only should contractual obligations be monitored, but so should the timing of contract formation activities. Owner management should review the schedule performance of the formation cycle by comparing planned and actual dates for release of the RFP, bid evaluation, and contract award in particular. An example contract formation schedule report focusing on these indicators is shown later in this chapter as Exhibit 23.

4. *Contract reports should allow "management by exception."* There is a sea of data that could be directed toward management at any given time. Much of it, it is hoped, demonstrates that things are going according to plan or that no corrective action is required. Many owners, however, feel as though they are "drowning in data but starving for information." *Data* give management the symptoms of contractual performance, and *information* allows them to determine the extent and expected outcome of problems—How bad is it, and will it get better or worse?—isolate the causes, apply corrective action, and forecast the ultimate impact

on the project. The difference between data and information can be demonstrated through an example. Suppose a $1 million contract is in effect for lump-sum work, that $900,000 has been charged the owner, and that approximately 50% of the work has been accomplished. Any number of conclusions may be drawn from these data. Some of them are:

The work is proceeding according to plan. Mobilization costs of $300,000 were paid, as stipulated in the contract, as soon as the contractor arrived at the site. Changed work in the amount of $200,000 has been ordered by the owner, and one-half of this has been completed ($100,000). Half of the original contract work has been completed ($500,000). Although the amount of mobilization costs may be subject to criticism, nothing indicates deviation from cost or schedule plans.

The work is proceeding according to plan. The contractor is actually 90% complete and has been paid for that amount. The 50% completion figure was calculated two days prior to the arrival of a large, contractor-furnished pressure vessel, for which he was paid $400,000.

The contractor's performance is unknown. The discrepancy between cost paid and progress to date is due to a difference between owner methods of (1) measuring progress in order to pay the contractor; and (2) measuring progress in order to determine project status. These two determinations were made by different owner personnel using different progress estimating criteria.

Errors exist in either the measurement of progress, contractor payments, or both.

The contractor has performed $400,000 worth of emergency changed work that has been authorized by site management but has yet to be reported to higher owner management. This amount has not been entered as a contractual commitment.

Because of a claim settlement caused by owner-ordered acceleration, the contractor has been awarded an additional $400,000 for overtime, extra crews, increased material costs, and labor inefficiency. This increase has not been reflected on a subsequent change order.

A major design change was made that caused the contractor to rip out installed material ($100,000) and reinstall it in another location ($300,000).

The list of possible reasons for such a discrepancy could go on and on. It would serve only to point out the need for "problem traceability" within a contract reporting system, as well as to demonstrate the care that must be exercised when planning for and accomplishing the collection of cost and performance data.

5. *Information should be timely.* Except from a "lessons-learned" standpoint, information that is available after the fact is without value. It does little good for owner management to be apprised of problems once their impact has been made. Of course, it is helpful to know why things went wrong, but it is much more important to be able to know that things are going wrong when it is happening. And management should also know, to the extent possible, that things may go wrong before they actually do. In order for this to happen, contract reports should be issued frequently and reflect timely information. This "early warning" allows owners to focus on problem areas before they get out of hand and to direct corrective action toward them as soon as possible.

## CONTRACT REPORTING HIERARCHY

In order to achieve reporting that allows management by exception, data traceability, and timely response to actual or potential problems, owners should consider a structured system of periodic reporting, such as that depicted in Figure 23. The documents presented therein show a recommended arrangement of (1) source documents; (2) contractual tracking and compliance documents helpful to the contract administrator in managing his work; (3) detailed contract reports covering schedule, claims, change orders, and backcharges; leading to (4) a summary report for higher owner management. The documents, records, and reports shown should be so structured that the data and information contained in each is integrated and that management, once it receives the summary report, is able to trace down the reporting system to understand the causes and effects of information it receives.

A monthly contract cost report that provides owner management with key information regarding contracting performance is shown as Exhibit 22. It incorporates and summarizes significant information shown on detailed contract reports, illustrated by Exhibit 23 (contract formation schedule report), Exhibit 24 (claims summary), Exhibit 25 (change order summary), and Exhibit 26 (backcharge summary). Each of these detailed reports is in turn fed by lower-level reports or source documents used for the contract transactions. These should be continually maintained and filed with active contract records. Several of these source documents have been described and illustrated in earlier chapters. Depending on the structure of the owner's project or company organization, detailed contract reports (change orders, backcharges, and so

**FIGURE 23.** Contract reporting hierarchy.

on) may also be submitted to management as part of a periodic contract or project report package.

## KEY PERFORMANCE INDICATORS

Regardless of the reporting system characteristics, certain "key indicators" are helpful when owner management is trying to understand contracting status and performance. Contract reports should be designed and implemented so that these are presented. Depending on the amount of contracting used and the elements of contract risk involved, most of the following key indicators should be monitored.

1.  Contract formation
      Number of qualified bidders
      Number of addenda issued

**EXHIBIT 22. Monthly Contract Cost Report**

## MONTHLY CONTRACT COST REPORT

| | |
|---|---|
| Number of Contracts Awarded | _____ |
| Change Orders Issued | _____ |
| Backcharges Issued | _____ |
| Unresolved Claims | _____ |

| | |
|---|---|
| Total Original Contract Price | \$ _____ |
| Authorized Price Increases | \$ _____ |
| Accepted Backcharge Amounts | \$ (_____) |
| Expected Claim Amounts | \$ _____ |
| Total Contract Forecasted Cost | \$ _____ |
| Current Budget Contract Costs | \$ _____ |
| Cost Variance Expected at Project Completion | \$ _____ |

Details of Contract Status:

| Contract No. | Description | Budgeted Cost | Original Committed Amount | Change Order Amount | Back-charge Amount | Present Committed Amount | Paid to Date | Estimated Cost at Completion |
|---|---|---|---|---|---|---|---|---|
| | | | | | | | | |
| | | | | | | | | |
| | | | | | | | | |
| | | | | | | | | |
| | | | | | | | | |
| | | | | | | | | |

**EXHIBIT 23.  Contract Formation Schedule Report**

## CONTRACT FORMATION SCHEDULE REPORT

| Contract No. | Description | | Specs and Drawings Received | RFP Issued | Prebid Meeting | Bids Received | Award Recommended | Contract Award | Commence Work |
|---|---|---|---|---|---|---|---|---|---|
| | | Scheduled Date | | | | | | | |
| | | Actual Date | | | | | | | |
| | | Scheduled Date | | | | | | | |
| | | Actual Date | | | | | | | |
| | | Scheduled Date | | | | | | | |
| | | Actual Date | | | | | | | |
| | | Scheduled Date | | | | | | | |
| | | Actual Date | | | | | | | |
| | | Scheduled Date | | | | | | | |
| | | Actual Date | | | | | | | |
| | | Scheduled Date | | | | | | | |
| | | Actual Date | | | | | | | |

**EXHIBIT 24. Claims Summary**

## CLAIMS SUMMARY

| Contract Number | Contract Description | Description of Claim | Claim Amount | Unresolved | Resolved | C.O. Amount | Remarks |
|---|---|---|---|---|---|---|---|
| | | | | | | | |
| | | | | | | | |
| | | | | | | | |
| | | | | | | | |
| | | | | | | | |

Project _____
Period Ending _____

| | This Period | To Date |
|---|---|---|
| Number of Claims Received | _____ | _____ |
| Total Amount Claimed | $ _____ | $ _____ |
| Number of Claims Resolved | _____ | _____ |
| Cost of Claims Resolved | $ _____ | $ _____ |
| Expected Cost of Unresolved Claims | $ _____ | $ _____ |

**EXHIBIT 25. Change Order Summary**

## CHANGE ORDER SUMMARY

| Contract Number | Description | Issued C.O.'s | | Pricing Method Used | | | Proposed/ Potential C.O.'s | | Original Contract Price | Amended Contract Price |
|---|---|---|---|---|---|---|---|---|---|---|
| | | No. | Amount | L.S. | U.P. | C + | No. | Amount | | |
| | | | | | | | | | | |
| | | | | | | | | | | |
| | | | | | | | | | | |
| | | | | | | | | | | |
| | | | | | | | | | | |
| | | | | | | | | | | |
| | | | | | | | | | | |
| | | | | | | | | | | |
| | | | | | | | | | | |

Project _____
Period Ending _____

| | This Period | To Date |
|---|---|---|
| Number of Change Orders Issued | _____ | _____ |
| Cost of Issued Change Orders | $ _____ | $ _____ |
| Number of Unresolved Change Orders | _____ | _____ |
| Expected Value of Unresolved Change Orders | $ _____ | $ _____ |

**EXHIBIT 26.  Backcharge Summary**

BACKCHARGE SUMMARY

| Contract Number | Description | Description of Backcharge | Backcharge | | | Unre-solved | Amount | C.O. No. | Remarks |
|---|---|---|---|---|---|---|---|---|---|
| | | | Cost | Amount | Accepted | | | | |
| | | | | | | | | | |
| | | | | | | | | | |
| | | | | | | | | | |
| | | | | | | | | | |
| | | | | | | | | | |

Project _____

Period Ending _____

| | This Period | To Date |
|---|---|---|
| Number of Backcharges Presented | _____ | _____ |
| Cost of Backcharges Presented | $ _____ | $ _____ |
| Number of Backcharges Accepted | _____ | _____ |
| Cost Recovered through Backcharges | $ _____ | $ _____ |
| Remaining Recoverable Cost | $ _____ | $ _____ |
| Remaining Unrecoverable Cost | $ _____ | $ _____ |

Number of responsive bidders, compared with number invited

Planned versus actual dates for formation activities

Frequency of use of letters of intent

Average effective period of letters of intent

Percentage of contract amounts priced as lump sum, unit price, and cost plus

Deviations from planned pricing methods

2. Contract administration

Planned payments versus actual payments

Number of change orders

Percentage of change order cost to original contract prices

Backcharges (number and amount)

Claims (number and amount)

Percentage of change orders priced as lump sum, unit price, and cost plus

Average elapsed time from identification of need for change, change approval, issuance of change order, and payment

Owner time charges to contract administration tasks

## REPORTING CHALLENGES

Despite the best efforts of some owners, contract reporting often falls short of its objectives. Certain problems persistently plague the design and implementation of effective contract reporting and should be recognized and eliminated. A partial list of the most common challenges to contract reporting follows:

1. Timeliness of information
2. Consistent cutoffs in collection of cost, schedule, and performance data
3. Integration of contract data elements, and integration of contract data with other project information
4. Data traceability from summary reports, through detailed reports, to source documents
5. Reporting of potential problems in addition to approved transactions, such as disclosure of pending claims and changes rather than of only those that have been resolved or approved

6. Exception reporting; that is, reporting of information deserving of analysis or futher management attention, and separating this information from the sea of data available

7. Reconciliation of initially estimated costs, such as when a cost-plus change order is issued, with eventual actual costs, when the changed work is completed and contractor charges are known

8. Management attention; that is getting owner management to evaluate and respond to information presented in contract reports.

# CHAPTER 20

# CONTRACT AUDITING

In order to determine the efficiency of contracting processes or to ensure that contract charges are correct, most owners consider the use of some sorts of contract audits during the course of a major construction project. Auditing differs from most of the other contract-related functions already described in that it is not directly required in order to conduct contracting business. Rather than being a specific set of activities that must be performed during the planning, formation, or administration of contracts, auditing represents an independent, arm's-length review of contracting processes as a whole or of specific transactions deserving particular management attention.

The purpose of each type of audit described in this chapter is to give management a view. In the case of cost audits, this view may be close and detailed—focused on specific charges to a specific contract. Operational audits, though, tend to give management a broader picture, a larger view of the contracting processes being undertaken. In any case, audits do not take the place of standard contract formation and administration activities and controls. They are performed in addition to the normal review and approval of invoices, time sheets, quantity-installed reports, acceptance testing, and all the other evidences of contractual compliance. These are all elements of the *execution* of contracting activities. The purpose of auditing is to verify that these are being executed as effectively as possible.

## TYPES OF AUDITS

There are three broad areas of auditing related to the management of construction contracts. Each is conducted differently with a different set of objectives.

1. *Financial Audits.* Financial audits are designed to ensure that financial information regarding the status of the owner or contractor represents actual conditions. Most large companies are audited yearly in order to provide the figures for financial statements regarding their fiscal position (assets, liabilities, debt, income, and so on). Usually performed by an independent auditing firm of certified public accountants, this type of audit leads to their attestation that the company's financial statements present fairly the financial position of the company and the results of its operations for the period. For most owners, this is a very limited review as far as contracting is concerned. In most cases, contracting charges and operations are beyond the scope of financial audits. Occasionally, because of the size or scope of an owner's or contractor's contracting effort, and therefore its impact upon financial position, financial audits do examine contracts to some extent. On the whole, however, these types of audits should not be relied upon to ensure that contracting is being performed correctly, or even to identify contracting deficiencies.

2. *Cost Audits.* Cost audits, on the other hand, have a direct application to many contracting elements. Their purpose is to test and verify charges made to the owner under a contractual agreement. As such, they involve detailed examinations of contractual billings. Their scope usually involves the matching of contractor charges to: (1) the contractual terms and prices; (2) verification that the work was performed; and (3) evidence of costs, labor hours, or other resource expenditures to achieve that performance. By its nature, cost auditing is performed after the fact; that is, amounts that have already been paid to a contractor are verified for accuracy, applicability, and reasonableness. Sometimes it is performed as a prerequisite to payment, but this is generally not the case.

3. *Operational Audits.* Operational auditing differs from the cost variety in that it examines *processes* that are being conducted or contemplated in the future. Operational audits look forward and seek to identify ways by which to improve contracting activities and controls for the benefit of the owner or contractor. Its goals are increased effectiveness, economy, and efficiency. Although costs are sometimes reviewed as indicators of contracting economy or effectiveness, operational audits focus their attention on control systems and processes.

## COST AUDITING OF CONTRACTS

Cost audits are performed by each contractual party when it feels that a separate check should be made of contractual charges, or when required under contractual terms and conditions. Their purpose is to verify charges, billings, payments, and/or receipts. They differ from the standard review that is made to verify the accuracy of charges when they are presented. This is a function of those responsible for making payments, usually construction accountants or the owner's accounts payable department. Cost audits provide an independent or second review of these amounts.

As such, cost audits should not be intended as a duplicative effort to that which is common business practice during the course of contract administration. Cost audits should represent more than just the checking of numbers that have already been checked or the review of invoices and statements that have already been reviewed. Instead, they should be targeted to specific high-risk areas. They should test charges or payments on a selective basis, and that selection should be guided by the degree of risk involved.

The conduct of all audits should be structured to a fairly well defined scope and work program. Cost audits are no different. Work programs that describe the step-by-step process of auditing, or the audit procedure, should be designed to identify errors and overcharges that are more apt to occur by virtue of the contractual pricing structure. Given the massive amount of data and records associated with payments under large contracts, it goes without saying that auditors do not have the time or personnel to review everything. The selection of charges to be tested is as important as the testing itself. For this reason, audit work programs should vary in scope and conduct, depending on the type of contract to which they will apply.

One would think that, particularly on large and sophisticated projects, this auditing would be planned and performed with great care. This is often not the case. Many contract audits, whether performed by an internal auditing group or an independent auditing firm, are merely superficial exercises. Typically, they center around a repetitive and mundane comparison of contractor invoices with associated cost substantiation and the contract itself. Much time and expense is spent chasing minor clerical or accounting errors through this process. This "shot in the dark" approach could be greatly enhanced if the audit detail and associated work programs were tailored to the risks inherent in each type of contract being audited.

It goes without saying that the risk of error, overcharging, and waste, from the owner's perspective, is greatest with cost-plus contracts and least with lump-sum contracts. Consequently, owner involvement and degree of audit detail should increase as the particular contract migrates toward a cost-plus agreement. But there are cost risks associated with any contract, regardless of its pricing structure. And each contract presents different elements of risk because of its particular pricing structure.

Let us briefly describe the areas of common error and overcharge for each major type of contract and just a few of the corresponding audit techniques applicable to each.

## LUMP-SUM (FIRM-FIXED-PRICE) CONTRACTS

For the lump-sum or firm-fixed-price contract, the area of frequent error or overcharge is with change orders. Applicable audit techniques that help avoid problems with change orders are:

1.  For selected contracts, verify that change orders represent changed or added work, that is, work not covered under the scope of the base contract or previously issued change orders for the contract under review or other project contracts.
2.  Where specific change order pricing provisions exist in the original contract, verify that they were used. If not, suitable justification for the deviation should be documented.
3.  Examine the time lag between identification of the need for extra work, such as a drawing revision, approval and issuance of the change order, and performance of extra work. Many change orders are issued after the fact, when the owner is in a poor bargaining position.
4.  Match lists of issued, pending, and "in-process" change orders provided by the owner with independent lists furnished by project contractors. Surprising discrepancies often exist.
5.  Examine the method by which change orders are priced; that is, was the most economical and controllable method chosen or are cost-plus terms being used more than necessary.

## UNIT-PRICE CONTRACTS

For unit-priced contracts, the area of frequent error or overcharge is in the determination of quantities. Applicable audit techniques that help avoid problems with determining quantities are:

1.  Match "quantity installed" reports with invoices.
2.  Check that following completion of the changed work *actual* quantity determinations were made for change orders that were based on estimated quantities when issued.

3. Investigate the need for independent verification of installed quantities through the use of as-built drawings, quantity surveys, and so on.

4. Verify that ancillary services and products normally included in unit prices, or specifically included per the contract, are not invoiced or paid for separately.

## COST-PLUS CONTRACTS

For cost-plus contracts, the area of frequent error or overcharge is in the determination and allowability of costs. Applicable audit techniques that help avoid these problems with costs are:

1. Examine controls to ensure that cost-plus resources, including cost-plus change orders, are not being used to perform fixed- or unit-priced activities by the same or other contractors—for example, double payment for resources.

2. Reconcile contractor time sheets with personnel brassing or other sign-in–sign-out records.

3. Verify allowability and reasonableness of general, administrative, and overhead charges.

4. Test allocation of contractor home office overhead to the project in accordance with contract terms.

5. Reconcile labor rates for construction craft personnel with actual local and national collective bargaining agreements in effect at the time the work was performed and/or contractor payroll records.

6. Check that construction equipment rental rates utilized were most beneficial to the owner; for example, monthly rates, rather than the more expensive weekly or hourly rates, used if equipment was rented for 30 days or more.

## OPERATIONAL AUDITING OF THE CONTRACTING PROCESS

Major contracting efforts entail a great deal of cost, time, and personnel. From an owner's viewpoint, the construction of a new facility often equals or exceeds the cost of any other concurrent operation. Most advocates of operational auditing feel that all major company operations should be periodically reviewed to assure company management that these operations are being performed with an acceptable level of control

THE ROLE OF AUDITORS

and to recommend improvements for the company's benefit. Operational audits, therefore, play an important role in the management of major contracts.

Operational auditing should determine: (1) if systems are in place to ensure an adequate level of control over each operation; and (2) if those systems are being operated correctly. Operational audits, therefore, begin with an examination of the controls that presently exist over each of the contract activities described in this volume. System deficiencies are noted and described. The second phase tests the operation of those systems that are in place. System and operating deficiencies are then noted, their negative impacts described, and the results reported to company or project management. Most operational audit reports include recommendations for both system and operating improvements that, if implemented by management, would reduce cost, liability, and inefficiency, as well as increase control over the processes themselves. The perspective of operational auditors, then, should be forward looking. They exist only to seek and recommend improvements, not to criticize operations or find fault with those responsible for inadequate operations.

Many medium and large corporations have instituted formal internal audit capabilities. Specific auditing departments or groups have been formed to conduct both cost and operational audits, and often these groups enjoy reporting relationships directly with top management. They operate independently of production or line management, serving as the "eyes and ears" of company executives. This organizational position and independence should allow internal auditors the freedom to seek out system and operating deficiencies and the unhampered ability to give top management an unbiased view of company activities and controls. It is not uncommon, however, for this position to be abused or underutilized. One of the greatest challenges of those charged with auditing responsibilities is to maintain their independence, their company-wide objective, and their working relationship with those responsible for elements under their review.

## THE ROLE OF AUDITORS

There are two distinct types of auditors that may perform the three major audits described earlier: internal and external. Internal auditors work for the company they are auditing, whereas external auditors are temporarily hired from the outside. External auditors, such as independent public accountants, are generally responsible for financial audits, although they may be engaged to audit costs or operations as well.

The theory of auditing, its value and use, is often well received by most responsible owners and contractors. Problems encountered with audits do not generally involve the need for an audit or the authority to have one conducted. They do, however, center around the role, conduct, and suitability of the auditors themselves. Financial auditing is usually so far removed from the daily activities, interest, or expertise of most people who deal with construction projects that it interferes little with their activities. Cost and operational audits, on the other hand, often bring the auditor into close proximity with those responsible for performing the functions that are the subject of the audit.

The most pervasive criticism of auditors, be they internal or external, is that they do not know the business of construction or understand the complexities of construction contracting. This is generally a well-deserved pronouncement and one that auditors and their management should heed. Much has been said about the need for credibility and understanding of construction processes on the part of the contract administrator. Auditors who will be involved with construction or contract activities share that need. The selection, training, and assignment of auditors is beyond the scope of this chapter. Let it suffice to point out that, particularly for operational auditors, actual functional experience together with formal training and exposure to the industry and its practices are essential. Provided auditors have the proper training and experience to perform their duties, their success depends upon the use of auditors and the willingness of company management to consider and act upon their recommendations.

One final recommendation is in order in this regard. Often owners station an auditor permanently on the construction site, expecting him or her to provide continuous monitoring of operations or costs. This is not recommended for several reasons. The primary one concerns the auditor's independence, or others' perception of that independence. Should he become deeply associated or involved with daily operations, or even become a part of ongoing execution tasks, this independence is lost. He would, in effect, be auditing work that is partially his. Another problem often encountered with this approach is that given the demands of contract administration or construction accounting responsibilities, the "independent" auditor is sometimes used to perform these activities. He often has more time to devote to his work than is required by the audit function, and a common tendency is to put him to useful work. When this happens, the audit presence and effectiveness are diminished.

## CONTRACT AUDIT CLAUSES

Most well-written construction contracts give the owner the right to inspect and audit the books of the contractor upon specified notice and subject to certain minor conditions. In addition, these contracts require the contractor to maintain accurate and complete cost records, particularly with cost-reimbursable agreements. It may be suggested that nonreimbursable contracts (lump sum or unit price) need not include such a clause. This could be true, providing that no instances requiring an audit or review of contractor costs arise. But there are many reasons why, even for lump-sum or unit-price contracts, auditing may be necessary to ensure that contractor charges are correct.

Typical claim situations are an example where this is the case. Even though the contractual setting may be fixed price, the very nature of claims is that they demand cost reimbursement for unforeseen events. Change orders that may be performed under cost-reimbursable conditions are other examples of the need for adequate audit rights in all contracts, regardless of their original pricing methods. Wise owners make sure that such clauses are included in their general conditions and are careful that, when inserted, they in fact allow the types of review or audit anticipated.

Even the most precisely worded audit clause may be subject to vastly differing interpretations when its use comes into play. Most clauses specify the right to have records audited but are not as specific as to who may perform this audit. If "company auditors" are specified, does this mean the owner's external public accounting firm, or the internal auditors employed by the company? The "records" that are subject to review are also difficult to identify unless specifically described in the audit clause. Questions sometimes arise as to which records are meant by "records," whether these are only contract- or project-specific contractor records, or whether they are *all* records maintained by the contractor, even those pertaining to other contracts for other owners. Again, proper and precise definition in the RFP and contract documents should be attempted to ensure that the owner's specific audit objectives are achieved.

Another issue that often surfaces when audit clauses are called upon concerns the right of the owner to audit the records of subcontractors or suppliers to the contractor. This right, if desired, should also be specified in what are called "flow-down" clauses—those that require the contractor to enforce similar clauses or obligations with his subtier agreements for the benefit of the owner.

## AUDIT AWARENESS

The frequency of internal and external audits of all types is increasing among construction projects. This is due in part to the growing number of interested parties involved in the effort. It is becoming more common to have more than one owner financing a project and more than one contractor performing work. Joint-venture agreements also increase the need for separate auditing to protect the interest of each partner. Closer outside scrutiny by the public, corporate stockholders, special-interest groups, and the like have increased the demands for account-ability of those responsible for construction contracting. All these factors have combined with economic and commercial pressures to expand both the role and frequency of construction audits, and this growth shows no sign of subsiding in the near future.

Because of the increasing scope and frequency of audit efforts, those responsible for contract management functions should evaluate each of their activities from a preaudit perspective. That is, they should plan and conduct their affairs with the expectation that they will be audited to one extent or another. This should cause a greater emphasis on thorough documentation of alternatives, decisions, and business trans-actions that occur during contract planning, formation, and adminis-tration. Forms, documents, records, and reports should be designed, used, and maintained to provide a clear audit trail. This "audit aware-ness" should not interfere with contracting activities by causing undue delay in decision making or by intimidating contracting personnel. On the contrary, it should achieve two interrelated objectives: (1) audit time and expenses will be reduced through clear audit trails, and the difficulties arising from the perception of lack of control will be eliminated; and (2) the knowledge that activities and transactions may be audited in the future serves to increase their efficiency and effectiveness for the immediate benefit of all concerned.

# GLOSSARY OF TERMS

ADDENDUM: Written or graphic instruments issued by the owner to bidders prior to the execution of the contract that modify or interpret the bidding documents (request for proposal), including drawings, by additions, deletions, clarifications, or corrections. These changes will become incorporated into the contract documents prior to award of the contract.

ADDITION TO CONTRACT PRICE: Amount added to the contract price by change order. See also Extra.

ADMINISTRATION, CONTRACT: The conduct of commercial transaction required for contractual compliance of a contract. Begins with contract award and ends with contract closeout or termination.

ADVERTISEMENT FOR BIDS: Published public notice soliciting bids for a construction project. Most frequently used to conform to legal requirements pertaining to projects constructed under public authority, and usually published in newspapers of general circulation in those districts from which the public funds are derived.

AGENT: One authorized by another to act in his or her stead or behalf.

AGREEMENT: (1) A meeting of minds. (2) A legally enforceable promise or promises between two or among several persons. (3) The specific contract document stating the essential terms of the construction contract and incorporating by reference the other contract documents.

AGREEMENT FORM: A document setting forth in printed form the general provisions of a contract, with spaces provided for insertion of specific data relating to a particular project. It is a part of the request for proposal sent to bidders.

ALTERNATE BID:   Change in project scope, proposed pricing method, or alternate materials and/or methods of construction offered by a bidder in lieu of, or in addition to, the solicited base bid.

APPLICATION FOR PAYMENT:   Contractor's written request for payment of amount due for completed portions of the work.

APPROVAL, ARCHITECT'S OR ENGINEER'S:   Architect's (engineer's) written or imprinted acknowledgment that materials, equipment, or methods of construction are acceptable for use in the work, or that a contractor's request or claim is valid.

ARBITRATION:   Method of settling claims or disputes between parties to a contract, rather than through litigation. An arbitrator or a panel of arbitrators, selected for their specialized knowledge in the field in question, hear the evidence and render a decision. Parties must agree to arbitrate, and arbitrators' rulings are usually enforceable by law.

ARCHITECT–ENGINEER (A–E):   An individual or firm offering professional services as both architect and engineer; used to refer to the company responsible for designing a facility.

AS-BUILT DRAWINGS:   Drawings of the contracted project, or portion thereof, showing the actual condition of the project elements at completion. They can be marked construction drawings, showing deviations from planned work, or separately drafted drawings. Submission of required as-built drawings is usually a prerequisite to a contractor's final payment.

AWARD:   A communication from an owner accepting a bid or negotiated proposal. An award is the final step in the contract formation process.

BACKCHARGE:   Process or document by which the owner charges a contractor or vendor for the cost of correcting work that the contractor or vendor failed to perform or performed incorrectly.

BASE BID:   Scope of work and corresponding amount of money stated in the bid as the price for which the bidder offers to perform the work. Usually refers to a bid that conforms to the RFP requirements.

BID:   A complete and properly signed proposal to perform the work or designated portion thereof for the prices stipulated therein. Supported by submittals required by the RFP. Also called proposal.

BID BOND:   A form of bid security executed by the bidder as principal and by a surety. The surety protects the owner from damages should the bidder fail to sign a contract to perform the work described in the RFP if the owner accepts its bid.

BID DATE:   The date established by the owner for the receipt of bids.

BID SECURITY:   The deposit of cash, certified check, cashier's check, bank draft, money order, or bid bond submitted with a bid and serving to guarantee to the owner that the bidder, if awarded the contract, will execute such contract in accordance with the terms of the RFP.

BOILER PLATE:   Term informally used to represent the collection of standard legal and commercial terms, conditions, and clauses used for all contracts for a project, such as general and special conditions.

BONDING CAPACITY:   The total dollar value of contracts that the surety will guarantee for a contractor.

BONUS AND PENALTY CLAUSE(S):   Provisions in the contract for payment of a bonus to the contractor for completing the work prior to a stipulated date, and/or for a charge against the contractor for failure to complete the work by such a date. These clauses are also used as incentives for specified cost and technical performance.

BRIEF-FORM (FIELD) CONTRACT:   Construction contract prepared and awarded by field office for short-term, low-cost work, not in the scope of present or planned construction contracts.

CERTIFICATE FOR PAYMENT:   A statement from the engineer, construction manager, or other owner's representative to the owner confirming the amount of money due the contractor for work accomplished.

CERTIFICATE OF INSURANCE:   A memorandum issued by an authorized representative of an insurance company stating the types, amounts, and effective dates of insurance in force for a contractor. The industry-wide "accord" form is generally used.

CHANGE ORDER:   A written order to the contractor signed by the owner or its agent (such as the A–E, or construction manager), issued after the execution of the contract, authorizing a change in the work or an adjustment in the contract price or the contract time. A change order may be signed by the owner's agent alone, provided the agent has authority from the owner for such procedure and that a copy of such written authority is furnished to the contractor upon request. A change order should also be signed by the contractor signifying its agreement to the adjustment in the contract price or the contract time. The contract price and the contract time should be changed only by change order.

CHANGES IN THE WORK:   Changes ordered by the owner consisting of additions, deletions, or other revisions within the general scope of the contract, the contract price, and the contract time being adjusted accordingly. All changes in the work should be authorized and documented by a change order.

CLAIM:   Request from a contractor for adjustment to the contract price, contract time, and/or contract requirements.

CLOSED SPECIFICATIONS:   Specifications stipulating the use of specific products or processes without provision for substitution.

CLOSEOUT:   The process of formally terminating a construction contract upon completion of the work or termination of the agreement because of other reasons (such as contract breach or default or cancellation of the contract).

CODES:   (1) Regulations, ordinances, or statutory requirements of a governmental unit relating to building construction and occupancy, adopted and administered for the protection of the public health, safety, and welfare. (2) Standards of design, construction, materials, tests, or other products or methods established by industry or governmental bodies and often cited or referenced in the specifications.

COMPLETION DATE:   The date established in the contract documents for substantial completion of the work.

CONDITIONS OF THE BID:   Conditions set forth in the instructions to bidders, the notice to bidders or advertisement for bids, the invitation to bidders, or other similar documents prescribing the conditions under which bids are to be prepared, executed, submitted, received, and accepted.

CONDITIONS OF THE CONTRACT:   Those portions of the contract documents that define, set forth, or relate to contract terminology; the rights and responsibilities of the contracting parties and of others involved in the work; requirements for safety and for compliance with laws and regulations; general procedures for the orderly prosecution and management of the work; payments to the contractor; and similar provisions of a general, nontechnical nature. The conditions of the contract typically include general conditions and special or supplementary conditions.

CONSTRUCTION MANAGER:   (1) Special management services performed by a contractor, the A–E, or others during the construction phase of the project, under separate or special agreement with the owner, or by the owner itself. Generally, this is not part of the contractor's or engineer's basic services but is an additional service sometimes included. (2) Designated individual serving as highest owner authority during the project construction period.

CONSTRUCTIVE CHANGES:   See Informal changes.

CONTRACT:   A legally enforceable promise or agreement between two or more parties.

CONTRACT DOCUMENTS:   The agreement, the conditions of the contract (general, special, and other conditions), the drawings, the specifications, all addenda issued prior to execution of the contract, all modifications thereto, and any other item specifically stipulated as being included in the contract documents.

CONTRACT PRICE:   The price stated in the owner–contractor agreement, which is the total amount payable by the owner to the contractor for the performance of the work under the contract documents.

CONTRACT TIME:   The period of time established in the contract documents within which the work must be completed.

CONTRACTOR:   (1) One who contracts. (2) In construction terminology, the person or organization responsible for performing the construction work and identified as such in the construction contract.

COST-PLUS-FEE AGREEMENT:   An agreement under which the contractor is reimbursed for its direct and indirect costs and, in addition, is paid a fee for its services. The fee is usually stated as a stipulated sum or as a percentage of cost.

CRITICAL PATH METHOD (CPM):   A planning and scheduling technique involving the charting of all events and operations to be encountered in completing a given process, rendered in a form permitting determination of the relative significance of each event and establishing the optimum sequence and duration of operations.

DATE OF COMMENCEMENT OF THE WORK:   The date established in a notice to proceed or, in the absence of such notice, the date of the agreement or such other date as may be established therein or by the contracting parties.

DATE OF SUBSTANTIAL COMPLETION:   The date when the work or a designated portion thereof is sufficiently complete, in accordance with the contract documents, so the owner may occupy the work or designated portion thereof for the use for which it is intended.

DEFAULT:   Contractor or owner's failure to perform under the contract.

DEPOSIT FOR BIDDING DOCUMENTS:   Monetary deposit required to obtain a set of request for proposal documents. Customarily, this is refunded to the bona fide bidders on return of the documents in good condition within a specified time.

DESIGN–BUILD:   Use of a single firm to both design and construct a facility.

DRAWINGS: The portion of the contract documents showing in graphic or pictorial form the design, location, and dimensions of the elements of the work to be constructed.

ENGINEER: Designation reserved, usually by law, for a person or organization professionally qualified and duly licensed to perform engineering services.

ESCALATION: A change in cost over time, either a decrease or increase, resulting from a change in the price of a particular element from the bid and/or contract price.

ESTIMATE, CONTRACTOR'S: A forecast, as opposed to a firm proposal, of construction cost, prepared by a contractor for a project or a portion thereof. (2) A term sometimes used to denote a contractor's application or request for a progress payment.

EXTRA: An item of work not included in the contractual scope, which may involve additional cost and time.

FAST TRACKING: (1) Method of design and construction whereby a series of tasks, jobs, or contracts are overlapped to reduce the length of time required to complete the project. (2) The overlapping of design and construction phases so that construction begins before design is completed.

FINAL ACCEPTANCE: The owner's acceptance of the work from the contractor when satisfied that it is complete and in accordance with the contract requirements. Final acceptance precedes final payment unless otherwise stipulated.

FINAL INSPECTION: Final review of the contractor's work by the architect, engineer, construction manager, or owner prior to final acceptance and final payment.

FINAL PAYMENT: Payment made by the owner to the contractor of the entire unpaid balance of the contract price as adjusted by change orders, backcharges, or other amendments.

FORCE ACCOUNT: Construction by the owner.

FORMAL CHANGES: Written directives from the owner to a contractor to change the time, cost, or scope of work from that originally specified by the contract documents.

FORMATION, CONTRACT: The series of tasks and activities leading to achievement of a written contract signed by both principal contracting parties.

FRONT-END LOADING: The practice by bidders of increasing the bid amount of items or work to be performed early and reducing the price of items to be performed later so as to maximize initial progress payments.

GENERAL CONDITIONS OF A CONSTRUCTION CONTRACT: General requirements that apply to all construction contracts for a particular project. The basic relationships between the contracting parties, general project rules, and commercial terms are contained in this section.

GENERAL CONTRACTOR: The function of the general contractor is to coordinate the activities of the various parties and agencies involved with the construction and to assume full, centralized responsibility for the delivery of the finished product within the specified time.

GUARANTEED MAXIMUM COST: Amount established in some cost-plus contracts as the maximum cost of performing specified work on the basis of cost of labor and materials plus overhead expense and profit. Unless modified by changes, the owner's payment obligation does not exceed this amount.

INFORMAL CHANGES: Changes to a contractor's scope of work, cost, or required time of performance brought about by acts of God, acts or omissions of the owner or others, or circumstances beyond the control of the contractor.

INVITATION TO BID: A portion of the bidding documents soliciting bids for a construction project for which bidding is not open to the public.

JOINT VENTURE: A collaborative undertaking by two or more persons or organizations for a specific project or projects, having the legal characteristics of a partnership.

LABOR AND MATERIAL PAYMENT BOND: A bond of the contractor in which a surety guarantees to the owner that the contractor will pay for labor and materials used in the formation of the contract. The claimants under the bond are defined as those having direct contracts with the contractor or any subcontractor.

LETTER OF INTENT: A letter or other communication issued upon award to the successful bidder, signifying an intention to enter into a formal agreement, usually setting forth the general terms of such agreement by referencing the request for proposal. Letters of intent are sometimes issued between contract award and signing of the conformed contract.

LIQUIDATED DAMAGES: Sum of money, usually given as dollars per day, stipulated in a construction contract as a penalty against the contractor for failure to complete the work by a designated time.

LUMP-SUM CONTRACT:   Contract in which a specific amount is set forth as the total payment for performance of the contract.

MECHANIC'S LIEN:   A lien on real property created by statute in all states in favor of persons supplying labor or materials for a building or structure for the value of labor or materials supplied by them. In some jurisdictions, a mechanic's lien also exists for the value of professional services. Clear title to the property cannot be obtained until the claim for the labor, materials, or professional services is settled.

MILESTONE:   (1) A key project event whose accomplishment is scheduled and against which progress is closely measured. (2) A specific achievement designated in a contract progress payment arrangement.

MULTIPLE PRIMES:   The use of several prime contractors, each under separate contract with the owner, to perform specific construction work.

NOTICE TO PROCEED:   Written communication issued by the owner to the contractor authorizing it to proceed with the work and establishing the date of commencement of the work.

NOTIFICATION:   Transmitted from the owner to a contractor informing it of anticipated changes in the work and requesting a response indicating the cost and schedule impact of the changes. Often serves as a transmittal for revised drawings or specifications.

PERFORMANCE BOND:   A bond of the contractor in which a surety guarantees to the owner that the work will be performed in accordance with the contract documents. Except where prohibited by statute, the performance bond is frequently combined with a labor and material payment bond.

PREQUALIFICATION OF PROSPECTIVE BIDDERS:   The process of investigating the qualifications of prospective bidders and selecting those who will be invited to bid, on the basis of their technical competence, financial integrity, and past performance.

PRIME CONTRACTOR:   Any contractor on a project having a contract directly with the owner (as opposed to a subcontractor).

PROGRESS PAYMENT:   Partial payment made during progress of the work on account of portions of the work completed. Usually made on a monthly basis.

PROJECT MANAGER:   The individual representing the owner who has overall responsibilities for a project.

PUNCH LIST:   A list of uncompleted or corrective items of work to be done to complete the contract work.

QUALIFIED BIDDER:   Contractor who has been allowed by the owner to bid when bidding is by invitation.

REIMBURSABLE EXPENSES:   Amounts expended for or on account of the project that, in accordance with the terms of the appropriate agreement, are to be reimbursed by the owner.

RELEASE OF LIEN:   Instrument executed by one supplying labor, materials, or professional services on a project that releases the individual's lien against the project property. Usually a prerequisite to final payment.

REQUEST FOR PROPOSAL (RFP):   The advertisement or invitation to bid, instructions to bidders, the bid (proposal) form, and the proposed contract documents, including any addenda issued prior to receipt of bids.

RESPONSIVE BID:   Bid submitted in good faith, properly signed, complete, and in prescribed form that meets the conditions of the bidding requirements.

RETAINAGE:   Also called retention. A sum withheld from progress payments to the contractor in accordance with the terms of the construction contract. Traditionally, it is 10% of payments due the contractor and payable upon final acceptance of the work.

RIPPLE EFFECT:   A change in one area, or to one contractor, that causes changes in other areas of the project or to other contractors.

SHOP DRAWINGS:   Drawings, diagrams, illustrations, schedules, performance charts, brochures, and other data prepared by the contractor or any subcontractor, manufacturer, supplier, or distributor that illustrate how specific portions of the work shall be fabricated and/or installed. They are submitted for the owner's approval prior to commencing the affected work.

SPECIFICATIONS:   The technical portion of the contract documents consisting of written descriptions of materials, equipment, construction methods, standards, and workmanship.

SPREAD SHEET:   Form used to record results of a bid evaluation.

SUBCONTRACT:   Agreement between a prime contractor and a subcontractor for a portion of the work for which the prime contractor is under contract with the owner to provide.

SUBCONTRACTOR:   A person or organization who has a direct contract with a prime contractor to perform a portion of the work.

SUB-SUBCONTRACTOR: A person or organization who has a contract with a subcontractor to perform a portion of the work.

SURETY BOND: A legal instrument under which one party (surety) agrees to answer to another party (owner) for the debt, default, or failure to perform of a third party (contractor).

TAKEOFF: Process whereby a cost estimator determines required resource quantities and applies applicable rates to obtain a cost estimate.

TIME AND MATERIALS (T AND M) CONTRACT: Contract that provides the contractor with reimbursement for all direct labor hours at a fixed hourly rate and for the actual costs of material used. The contractor's overhead and profit are embedded in the hourly rate.

TURNKEY CONTRACTOR: A contractor responsible for all phases necessary to complete design and construction of a facility. Theoretically, once the owner has expressed its needs, the turnkey contractor does everything up to and including turning over the facility's "key" to the owner.

UNBALANCED BID: A deliberate practice by bidders of weighing or factoring component bid amounts to: (1) front-end load; or (2) increase unit prices for additions that the bidder expects to occur in significant quantities or decrease unit prices for items that the bidder feels will not be ordered in large quantities or may later be deleted from the scope of work.

VENDOR: Supplier of material, equipment, or finished goods to be used in construction without labor and supervision for their installation or erection.

WAIVER OF LIEN: An instrument by which a person or organization who has or may have a right of lien against the property of another relinquishes such right. In a few jurisdictions, this right can be waived even before the construction required by the contract documents.

WORKMEN'S COMPENSATION INSURANCE: Insurance covering liability of an employer to his employees for compensation and other benefits required by workmen's compensation laws with respect to injury, sickness, disease, or death arising from their employment. Required by law in each of the United States.

# INDEX